In the Beginning...

In the Beginning...

*A Scientist
Shows Why
the Creationists
Are Wrong*

CHRIS McGOWAN

Prometheus Books
Buffalo, New York

Published 1984 by
Prometheus Books
700 E. Amherst Street
Buffalo, New York 14215

Printed in the United States of America

ISBN 0-87975-240-8
Library of Congress Catalog Card No. 83-62997

Dedicated to the memory of Tom Fairley – a man who respected the truth.

Contents

Acknowledgements

MY PRESENT INVOLVEMENT in the creation controversy dates to early 1981 when Douglas
Cunningham, Science Head of the Bruce Penin-
sula District School, invited my participation
in a seminar at his school. I thank him for
alerting me to the seriousness of the problem
and salute his dedication to scholarship and to his students.

I thank Pamela Purves for typing, and Dr. Rosemary Johnson
and Jeff Thomason for reading the manuscript and making
many valuable suggestions.

Most of the illustrations were drawn by Anker Odum and
Jay Sobel and I thank them both for their excellent work.
The magnificent photographs of primates were taken by Bill
Robertson; Brian Boyle, of the Royal Ontario Museum Department of Photography, took the author photograph and did
most of the copying, and Allan McColl, also of the ROM, did
most of the printing.

Liz McGowan read proofs, corrected errors, made suggestions,
gave encouragement, and put up with my night-long writing
vigils with characteristic good nature. I thank her for all this
and much more besides.

For permission to reproduce certain illustrations in this book
I thank the Royal Ontario Museum and the United States Geo-

logical Survey. I also thank the Ontario Arts Council for a grant which has been used to offset the cost of some of the illustrations.

I learned a good deal about the creationists' viewpoint from my discussions with Ian Taylor and the Reverend Paul Vaughan. We will always disagree, but continue to enjoy cordial relationships and partake in good-humored exchanges of ideas. Dr. H. M. Morris, who has come under such attack in this book, generously gave me permission to take quotations from his own work. I thank him for this and hope I have fulfilled his wish not to be quoted out of context. Dr. D. T. Gish has also come under attack here, and I hope that I have expressed his views fairly.

I thank Father J. A. Weisheipl of the Pontifical Institute for Medieval Studies for his interest in this issue and for several useful discussions on theology. I am also grateful to my colleagues at the Royal Ontario Museum and at the University of Toronto for their valuable discussions and much-appreciated support.

My thanks to Dr. R. L. Zusi of the Smithsonian Institution for the loan of hoatzins and to Arlene Reiss of the Department of Vertebrate Palaeontology, Royal Ontario Museum, for X-raying them.

I would also like to thank the Natural Sciences and Engineering Research Council of Canada for their generous support of my research program over the years.

The decision to write this book was arrived at in consultation with the late Tom Fairley, who also introduced me to Macmillan of Canada. His advice was characteristically sound.

Working with Macmillan of Canada has been a most rewarding experience and I am especially grateful to Anne Holloway. I thank her for the meticulous care she took in editing the manuscript, for the efficient way she guided it through the various editorial stages, and for her good advice and wise counsel (American spelling notwithstanding).

Preface

Nobody living in the developed world can doubt that they are in the midst of a scientific and technological revolution. Space exploration and genetic engineering, home computers and satellite television, are just some of its manifestations, and none of these could have come into existence without scientists. But scientists have to be trained, and this training begins in the classroom.

When America found herself beaten into space by the 1957 launching of Sputnik 1, a commitment was made to radically upgrade the quality of this training. Today the United States has taken the lead. This is documented by the fact that during the past decade American scientists have won half again as many Nobel Prizes as all other scientists combined. If the United States is to retain this staggering lead, she must provide the next generation of scientists and technologists with the best possible education. But this ideal is being threatened by the current upsurge of Creation Science. So far the Creation Scientists—those who claim scientific support for their belief in the biblical account of creation—have concentrated their attack on evolutionary biology, but this is only the thin end of the wedge. In order to accommodate their beliefs in a 10,000-year-old earth, the Noachian flood, and all the rest, they have to distort other

branches of science out of all proportion. The resulting con-
coction is *not* science, but good old-fashioned fundamentalist
religion, all spruced up with scientific terminology and ideas to
look like science. Little wonder that none of their arguments
stand up to close inspection, as we shall see in the following
pages.

As part of their pressure sales technique, they allege that if
you are not for them you must be against them, and therefore
against God and Christianity. What is clear from reading their
literature and attending their debates is that they do not represent mainstream Christianity, and that they are as unorthodox
in their theology as they are in their science.

The leading creationist spokesmen are prolific writers who
touch on all aspects of science: astronomy, nuclear physics,
geology, ecology—you name it, they write about it, though
usually with an appalling disregard for scientific accuracy. They
tackle the origin of the universe and the origin of life and the
development of species with equal aplomb—amoeba to man, as
they call it.

I do not claim to be as talented or as widely informed, so I
am going to stick largely to my own subject. I do not discuss
the origins of the universe, which is best left to astronomers,
and I deal with the origin of life only in fairly general terms.
What this book will outline, however, is the subject of organic
evolution, how new organisms have come into existence. Much
of my evidence is based on fossils and the rocks in which they
are found, because the fossil record is the most concrete source
of documentation for the slow process of evolution.

The book is primarily written for open-minded people who
want to be able to see the creation issue in its true light. I hope
that any teachers facing the battle over creationism in their
classrooms will read this book. The reason for this is simple: if
our school boards, or, worse still, our education departments or
ministries, are unwise enough to allow themselves to be duped
by the creationists, it will be our high school teachers and their
pupils who will be in the front line. Most of my academic
colleagues do not think this could ever happen, but I am less
convinced. Politicians respond to pressure, and the creationists,
with their petitions, their leaflets, and their accusations of anti-
Christian attitudes among their opponents, are generating a fair
head of steam.

I have included many quotations from the creationists in the
text. The reason for this is twofold: first, it removes any chance
of my misrepresenting their views, and, second, many of the

things creationists say are so incredible that the reader might suspect me of embellisment if I paraphrased them.

This book was not written in the hope of converting any creationists. I believe their minds are already made up. It was written to show just how groundless their arguments are and what a backward step it would be to heed them in the guise of "fairness." I hope this book will steer some people away from the creationist cause, while giving others a little ammunition with which to fight back.

CHRISTOPHER MCGOWAN

Curator-in-Charge
Department of Vertebrate Palaeontology
Royal Ontario Museum, and
Associate Professor, Department of Zoology
University of Toronto, Toronto, Canada

ONE

Science, Non-Science, and Nonsense

 MY JOB AT TORONTO'S ROYAL ONTARIO MUSEUM gets me into some pretty unusual situations — a monastery in Kremsmünster to examine fossil reptiles, a rain forest in Queensland to study flightless birds, an extinct volcano in Galapagos to see giant tortoises — but none of these experiences can compare with sitting in a darkened room listening to a lecture on Creation Science.

A spectator at one of these remarkable events could be forgiven for believing he has been whisked back in time to some dark and unenlightened age, when man's inquiring mind was not permitted to stray beyond the bounds of the Old Testament. The creationists, you see, believe that the book of Genesis is a precise account of the origins and early history of our living world. God literally created the universe, and the world, and all its inhabitants, in a six-day period. Adam, the first man, was created from dust, and Eve, the first woman, was made from one of Adam's ribs. Several generations later came the Great Flood, completely covering the earth and extinguishing all creatures, save those that Noah took into his ark.

People who share this literal belief in the Bible, usually referred to as fundamentalists, are very much a minority group. The majority of Christians regard the book of Genesis as a collection of parables which illustrate the point that there is a

God, a God who has dominion over the world. In the same way most historians probably believe that there was an early English king called Arthur, but I doubt whether any of them believe in Merlin, or dragons, or the other trappings of the Arthurian legend.

The minority status of the creationists was most strikingly demonstrated to me during my first formal encounter with them, at a two-day seminar on the creation-evolution controversy at the Bruce Peninsular District Secondary School, in a rural area of Ontario. In my talk, which followed the creationist's presentation, I outlined my position as an evolutionist. I explained that I believed present-day animals and plants were descended from earlier forms. Birds, for example, are the modified descendants of reptiles. Although the process is too slow to be seen during our own short lives, we have adequate documentation of evolution in the fossil record. At the conclusion of my talk the session was opened up for questions, and I was a trifle taken aback to see so many hands raised by the clergy in the audience. Had I ruffled their ecclesiastical feathers with my discussions of *Archaeopteryx*, the transitional bird? Not a bit! The clergymen were on my side, on the side of evolution. At first I was a little surprised, but I should not have been, because a belief in God is not mutually exclusive of a belief in evolution. The creationists, of course, vehemently disagree that these two philosophies can coexist, and one of my major objections to the movement is the pernicious way they present creation and Christianity as one side of a coin, with evolution and atheism on the other. The issue is *not* one between Christians and evolutionists, but between a vociferous Christian minority — the creationists — and evolutionists.

My other major objection is the way creationists misinterpret and distort scientific evidence to support their religious beliefs. Their potpourri of misinformation, misunderstanding, and plain ignorance is then given the veneer of respectability with the term "Creation Science." But why should they want to label their religious beliefs as a science? The answer is very simple: they want to introduce their unorthodox views into the classroom. In the United States, birthplace of the Creation Science movement, religion is not permitted to be taught in the schools. By describing their religious views as science, though, the creationists can lobby for equal time in the science classroom to teach creation alongside evolution. They have been quite successful, and many school boards in the United States have capitulated to their persuasion. What, after all, could be more reasonable than giving equal time to an alternative scientific theory for the origins of new species? At this point, we need to explain what is meant by science and the scientific method.

For some people science is something incomprehensible that people in white coats do in laboratories, but however complex some aspects of science may be, the scientific method is perfectly simple. A scientist makes observations, and from these observations he gets some idea of what is going on — he constructs an hypothesis. The hypothesis is then tested, often by setting up experiments to see whether it holds true. If the hypothesis passes the tests, especially if these are many and varied, there is good indication that the hypothesis is true. For example, suppose a scientist observes that anglers fishing in a particular lake start catching fewer fish after a new pulp processing plant starts operating in the vicinity. A reasonable hypothesis is that the pulp plant is damaging the lake environment, thereby reducing fish populations. A number of different types of tests could be devised to see if the hypothesis holds true:

1. Angling records are checked for this particular lake to see whether similar declines have been recorded in the past. No such declines have been recorded.
2. Angling records are checked for other lakes in the vicinity of the plant. Similar declines in fish populations are recorded, all commencing at the same time.
3. Water samples are removed from the affected lakes and analyzed for pollutants. Chemicals are detected that are known to be used in the pulping process.

None of these tests have shown the hypothesis to be false; indeed they all strengthen it, and there is probably sufficient evidence to take the plant operators to court. But this does not mean that the hypothesis is necessarily true; all it means is that it has not yet been shown to be false. Can you think of further tests that might falsify the hypothesis? Suppose a survey is now made of the anglers fishing the affected lakes. The survey shows that before the plant opened, only a few local inhabitants fished the lakes; afterwards, several hundred plant workers fished the lakes. The hypothesis is weakened by the test because the decline in fish populations might well be due to overfishing. Suppose the pollutants found in the affected lakes are tested on healthy fish in the laboratory and found to be harmless? The hypothesis is rejected. If the case has already gone to court, these new findings would undoubtedly win an acquittal for the plant operators.

Hypotheses that stand up to rigorous testing without being falsified are usually referred to as theories. There are numerous scientific theories: the atomic theory, the theory of relativity, the wave theory of light, the theory of continental drift, and the theory of evolution, to name but a few. The creationists would disagree with me that the last-named is in fact a legitimate theory, but before we falsify their argument we need to identify an authoritative spokesman for their views, and it is easiest to do this if we know something about the creationists.

A history of the Creation Science movement is out of place here, but suffice it to say that its roots lie in California, and that the Institute for Creation Research and the Christian Heritage College, both in San Diego, are its principal seminaries. The Institute publishes a monthly periodical, *Acts and Facts*, and the San Diego-based Creation Life Publishers have published a number of books on creation. Most significant among these are *Evolution: The Fossils Say No!* by Dr. D. T. Gish, associate director of the Institute for Creation Research, and *Scientific Creationism* by Dr. H. M. Morris, the Institute's director. (It is worth noting here that Dr. Gish has a Ph.D. in biochemistry from the University of California [Berkeley], and that Dr. Morris has a Ph.D. in hydraulics and hydrology from the University of Minnesota.) The latter book, written in consultation with the scientific staff and technical advisory board of the Institute, sets out the creationist case in great detail, and it is generally regarded by creationists as being the authoritative work on the subject. I have drawn freely from the books of Drs. Morris and Gish in setting out the creationists' case.

Whatever our differences may be, evolutionists and creationists are agreed on one thing: the meaning of science. Dr. Gish, for example, recognizes that science involves observation and testing. He also recognizes that since creation can be neither observed nor tested, it cannot be considered as scientific. Here again I find myself in complete agreement. But the next step in his argument is unacceptable: evolution is rejected as a scientific theory on the grounds that it has not been observed — no one has ever seen the origin of a new species. Neither has anyone seen an electron, or any other fundamental particle, but this does not invalidate the atomic theory. Let there be no doubt that the theory of evolution, that is, the theory that living organisms have descended, with modification, from earlier inhabitants of the earth, is a scientific theory in every sense of the word. As we will see in Chapter 2, it was through observation that Charles Darwin was led to believe that evolution rather than creation had occurred. Since Darwin's time the theory has been subjected to extensive testing in many different fields, and none of these tests have falsified the theory. We will be examining some of these tests in later chapters, but it would be useful to mention some of them now.

The theory of evolution would be falsified if we found that the earliest fossils were not the simplest, or that all the different types of organisms appeared at the same time. The fossil record supports evolutionary theory on both of these grounds. If the theory of evolution is true, we would expect to find biochemical as well as anatomical similarities between organisms that we believe to be closely related. This is precisely what we find; human and chimpanzee proteins, for example, have been found to be almost identical in structure. We can also expect the geographic distribution of organisms to support the theory of evolution; if organisms have evolved from earlier organisms, we should expect closely related species to be found in close proximity, especially those with limited powers of dispersal. This is precisely what we find; kangaroos, for example, are found only in Australia, hummingbirds in the Americas, Darwin's finches only in the Galapagos archipelago. The restriction of certain organisms to certain geographic regions is called endemism, and endemic species are difficult to reconcile with the story of Noah's ark (see Chapter 5).

However passionately creationists may argue their case, evolution *is* science and creation "science" is *not* science. As such, creationism has no more place in the biology classroom than pre-Galilean astronomy has in astronomy classes, alchemy has in the chemistry lab, or discussions of body humors have in medical schools. Our discussion could end right here, but we will play along with the creationists and see how they develop their case.

Having dismissed the theory of evolution as being non-scientific, creationists place creation and evolution side by side as equally plausible "models" for the explanation of the living world — neither one being any more scientific than the other. We notice, however, that they continue to refer to their subject as creation *science*. They then attempt to discredit the "evolution model," thus leaving the "creation model" as the only viable alternative. This specious line of argument, which assumes that there are only two alternatives, is very popular with politicians. If a powerful enough attack is made upon the party in power, the opposition party automatically looks good!

Having cast evolution and creation in the same mold, creationists consider the implications of the two models. The following quotation from Dr. Morris illustrates certain anomalies in the creationists' position:

> The completed original creation was perfect and has since been "running down."
>
> The creation model, thus postulates a period of special creation in the beginning, during which all the basic laws and categories of nature, including the major kinds of plants and animals, as well as man, were brought into existence by special creative and integrative processes which are no longer in operation. Once the creation was finished, these processes of *creation* were replaced by processes of *conservation*, which were designed by the Creator to sustain and maintain the basic system He had created.
>
> In addition to the primary concept of a completed creation followed by conservation, the creation model proposes a basic principle of disintegration now at work in nature (since any significant change in a *perfect* primeval creation must be in the direction of imperfection). Also, the evidence in the earth's crust of past physical convulsions seems to warrant inclusion of post-creation global catastrophism in the model.

Notice that although the creation is described as perfect, and even though it is followed by a phase of conservation, it is said to be disintegrating. This is surely contradictory, because if the creation were initially perfect there would be no need for conservation. Furthermore, if there *were* conservation, there should be no disintegration. In spite of this unresolved contradiction, we can see that the main point of Dr. Morris's argument is that the creation model predicts decline — change from organization to disorganization, from complex to simple. This prediction, the creationists point out, contrasts with the evolution model, which predicts a progression, from disorganization to organization, from simple to complex.

In order to support their view of a disintegrating world, creationists invoke the Second Law of Thermodynamics, which applies to non-living systems and states that these systems tend to change to become more random, less complex. This strategy is very effective; baffle people with science and they invariably bow to your superior knowledge. We will digress for the moment and discuss thermodynamics in some detail, and for good reasons. First, it is an important plank in the creationist platform; secondly, it is a subject that most of us are unfamiliar with; and thirdly, it serves as an excellent example of how

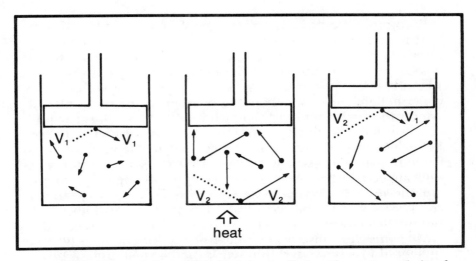

LEFT: *Molecules moving randomly in a closed cylinder. The average velocity of the molecules, V_1, is unchanged after collision with the walls.*

MIDDLE: *Heat increases the average velocity of the molecules to V_2.*

RIGHT: *The piston is allowed to move upwards until the average velocity of the molecules is V_1 again. In slowing down, the molecules transfer kinetic energy to the piston.*

creationists bring totally irrelevant pieces of science to the support of their beliefs.

The word thermodynamics, combining the Greek words for heat and motion, is concerned with the relationships between heat and the motion of molecules. Having to do with the behavior of molecules, particularly with gas molecules moving in confined spaces (closed systems), thermodynamics is a rather abstract subject, but it is easy enough to understand the basics. Visualize a cylinder fitted with a piston and containing a gas. The piston is kept stationary at the moment; therefore the volume of the space occupied by the gas molecules is constant. There are, of course, billions of gas molecules in the container, each one moving randomly within the space, colliding with other molecules and colliding with the walls of the container. The molecules are not all traveling at the same velocity; some are traveling faster than others, and their velocities change each time they collide with another molecule, sometimes increasing, sometimes decreasing. A similar situation might pertain in a crowded dance hall.

Each molecule has energy, energy of motion, or kinetic energy to give it the correct term. The kinetic energy of a moving object is the product of half its mass and the square of its velocity; fast, light objects, like bullets, and heavy, slow objects, like oil tankers, have large kinetic energies. The walls of the container are continuously being bombarded by molecules. If we could watch a molecule colliding with the wall, we would observe that its velocity after impact was the same as it was before. As the velocity of the molecule is unchanged (only its direction is changed), it follows that its kinetic energy is unchanged. Such collisions, involving no changes in energy, are said to be perfectly elastic. If I ran against a brick wall my collision would not be very elastic. I might bounce off the wall, but my velocity after impact would be considerably reduced. I would therefore lose a great deal of my kinetic energy, and this would be converted into work (perhaps enough to break bones), some heat, and a fair amount of sound. A billiard ball's collision with the cushion of a billiard table is far more elastic, and the velocity (hence kinetic energy) after collision is not greatly reduced.

The molecules do not lose any kinetic energy to the walls of the container. Furthermore, any loss of kinetic energy from one molecule through collision with another is completely transferred to the second molecule. Consequently the sum of the kinetic energies of all the molecules, called the internal energy of the system, remains constant.

If we now heated the container, the molecules would speed up, increasing their individual kinetic energies and thereby increasing the internal energy of the system. The increase in internal energy, dE, is equal to the amount of heat energy, dQ, transferred to the container: $dE = dQ$. This increase in internal energy can be used to perform work. If the piston is allowed to move as the molecules strike it, their additional kinetic energy can be completely transferred to the piston. Suppose the average velocity of the molecules prior to heating was V_1, and after heating was V_2. If the piston is allowed to move back until the average velocity of the molecules is V_1 again, the internal energy of the system is restored to its original value. The amount of work, dW, done on the piston is equal to the amount of heat energy added to the system: $dW = dQ$. If the piston were only allowed to rise up part of the way, the work output would be less than the heat input; hence the internal energy would be higher than it was initially.

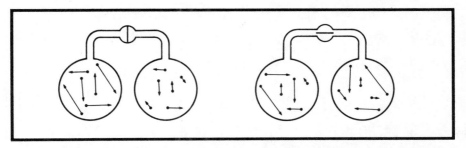

LEFT: *The gas in the left-hand flask is hotter than that in the flask on the right; therefore its molecules are traveling faster. The closed tap prevents mixing.*

RIGHT: *When the tap is opened, the molecules distribute themselves randomly between the two flasks such that the gas temperature in the two flasks is equal.*

Instead of heating the container, we might have chosen to push the piston down. This would have imparted additional kinetic energy to the molecules, similarly effecting an increase in the internal energy of the system. Conversely, pulling the piston up would have reduced the internal energy of the system. The relationship between heat, internal energy, and work is expressed by the equation:

$$dQ \quad = \quad dE \quad + \quad dW$$
$$\text{(heat)} \quad \text{(internal energy)} \quad \text{(work)}$$

This is the First Law of Thermodynamics. Expressed in words, this law says that the total energy in a closed system is constant.

The First Law of Thermodynamics, then, concerns the conservation of energy within a closed system, and, as we shall see in the following example, it is of limited use. Suppose we have two sealed flasks, of equal volume, one filled with a hot gas, the other with a colder one. The two flasks are joined together by a pipe but prevented from mixing for the moment by a closed tap. What is the likely outcome when the tap is opened? All that the First Law of Thermodynamics tells us is that the total energy of the system will remain the same. If E_1 is the internal energy of the hot gas and E_2 that of the cold gas, the total energy after opening the tap remains at $E_1 + E_2$. The first law tells us nothing about the individual behavior of the two groups of molecules. Will they stay in their respective flasks or will they mingle? The first law does not rule out the possibility that the hot gas will reach a higher final temperature and the cold gas a lower one, though this seems unlikely.

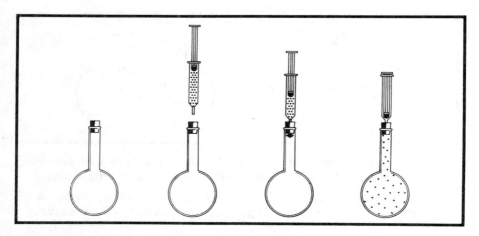

Gas molecules always distribute themselves to occupy all available space.

What actually happens when the tap is opened is that the two gases rapidly mingle. Collisions between the molecules result in an increase in the average kinetic energies of the (initially) cold molecules, and a corresponding decrease in the average kinetic energies of the others. The final temperature of the gas mixture is intermediate between the two initial temperatures. The outcome of this experiment, and others like it, is contained in the Second Law of Thermodynamics. This says that if an isolated system, free from external influences, is initially in a state of relative order, it passes into relative disorder, eventually reaching a state of maximum disorder. In this experiment the system passes from a state where two groups of gas molecules are ordered into two separate regions, to a state of disorder where the molecules are all jumbled up. We can illustrate this law with a second experiment. Suppose we take a flask, suck all the air out of it so that we have a vacuum inside, then close it with a rubber bung. A hypodermic syringe is then used to inject a small quantity of gas into the flask. Initially the gas molecules are ordered in a small cloud around the hypodermic needle; the next instant they have distributed themselves evenly throughout the flask. They have become less organized, more disorganized. The term entropy is used to indicate the degree of disorder in a system. Systems that have become more disorganized are said to have an increased entropy; the entropy of an isolated system, free from external influences, tends to increase to a maximum value. We must emphasize that the laws of thermodynamics apply *only to closed systems*, and are concerned with the behavior of molecules.

What, you may well ask, has all this got to do with the appearance of new organisms on the earth? Absolutely nothing! The origin of life, and the evolution of progressively more complex organisms, took place on the earth's surface, and this cannot possibly be described as operating in a closed system. Energy is free to flow in and out of the system, and the laws of thermodynamics are therefore absolutely irrelevant. Dr. Morris attempts to brush this objection aside by saying that "Although it is true that the two laws of thermodynamics are defined in terms of isolated systems, it is also true that in the real world there is no such thing as an isolated system." This is an illogical argument because the laws of thermodynamics specifically pertain to closed systems. Creationists can wriggle as hard as they please but they cannot alter the fact that the laws of thermodynamics are irrelevant to the discussion of evolution.

To return to Dr. Morris's statement about the creation model, there are two more points we should consider. First, by stating that basic laws were "created," he implies that the "evolutionary model" predicts that the basic laws have evolved. Surely the creationists are not seriously saying that we, the evolutionists, believe that the basic physical laws — laws of motion, gravitation, and the rest — evolved? But this is the conclusion he reaches a few pages further on:

> the law of gravity, the laws of thermodynamics, the laws of motion, and all other truly basic laws have apparently always functioned in just the way they do now, contrary to a prediction of the basic evolution model.

I have yet to encounter an evolutionist who has ever suggested that the physical laws might have evolved.

The last point I want to make regarding Dr. Morris's statement is the introduction of the concept of a global catastrophe into the creation model. When Dr. Morris refers to evidence for past physical convulsions in the earth's crust, what particular evidence has he in mind? One might anticipate a discussion of the evidence for earth upheavals later in his book (perhaps universal earthquakes, volcanoes, or the like), but this is not the case. His main thrust is not to produce evidence for global catastrophe, but to attack the opposite view: uniformitarianism. This term, sounding like some political doctrine, dates back to the early part of the last century. It was first used to describe the idea that the geological events of the past occurred relatively slowly and steadily, by "ordinary" processes, processes that were explicable in terms of science, rather than by global catastrophes caused by supernatural forces. Lyell, whose great work *Principles of Geology* influenced Darwin so much, extended the meaning of the word uniformitarianism to include the concept that geological phenomena of the past can be explained in terms of processes operating today. Deep valleys, for example, can be explained by the cutting action of running water — the process is very slow, but, extrapolated over countless years, it can cause major changes. Not all geological processes occur slowly and sedately like this, though; volcanoes, landslips, flash floods, and the like effect rapid changes. Such events, which often wreak great havoc and suffering, are popularly described as catastrophes, but we have to make the distinction that they are always localized rather than global, and that they are not caused by supernatural phenomena.

Dr. Morris attacks uniformitarianism, using the standard creationist strategy of quoting various criticisms made by (orthodox) scientists. Dr. Morris argues that since the principle of uniformitarianism is under such serious question by geologists, catastrophe is the only viable alternative. While it is true that many geologists have criticized uniformitarianism, it is important to know what criticisms have been made. One of the major criticisms pertains to the gradualistic implications of the term uniformitarianism, emphasizing as it does uniform rates of change. As we pointed out earlier, geological processes often involve rapid changes rather than slow, uniform rates of change. What is in question here, then, is not the concept that past geological processes can be explained in terms of process occurring today, but the gradualistic implications of the term uniformitarianism. The British geologists A. and D. C. Holmes make the point that the difficulty is removed if the term *actualism* (favored by European geologists) is used instead of uniformitarianism, emphasis being placed on the concept rather than on the rate. A second criticism that has been made of uniformitarianism is that there are some geological phenomena that are not easily explained in terms of processes occurring today. This does not mean, of course, that these phenomena are best explained in terms of supernatural processes.

Geologists have clearly not abandoned the principle that geological phenomena can be explained in terms of processes occurring today. All they are pointing out is that these processes are sometimes rapid rather than gradualistic, and that some of the geological phenomena cannot be satisfactorily explained in terms of present-day processes. This position is poles apart from Dr. Morris's assertion that all geological phenomena are best rationalized in terms of unique global catastrophes.

Why do creationists want to establish the existence of a global catastrophe? Dr. Morris tell us that

> The creation model is fundamentally catastrophic because it says that present laws and processes are *not* sufficient to explain the phenomena found in the present world.

He goes on to argue:

> Although some [paleontologists] have visualized fossilization as a slow process, brought about by gradual application of heat, pressure, chemical replacement, etc., it should be obvious that the actual formation of potential fossils in the first place, before other processes can start to work on them at all, requires rapid and compact burial of the organisms concerned, and this requires catastrophism.

Dr. Morris develops his theme of catastrophism to include the formation of the rocks that enclose the fossils (generally referred to as the geological column).

> The creation model, on the other hand, must interpret the [geological] column in terms of essentially continuous deposition, all accomplished in a relatively short time — not instantaneously, of course, but over a period of months or years, rather than millions of years. In effect, this means that organisms represented in the fossil record must all have been living contemporaneously, rather than scattered in separate time-frames over hundreds of millions of years. . . . there is no reason to doubt that man lived at the same time as the dinosaurs and trilobites.

Now we have the answer to our question. Creationists *must* argue that fossils were once living contemporaneously, and that they were formed and laid down in the rocks in a very short time (catastrophically); otherwise they have the embarrassment of trying to explain why the earliest fossils are the simplest, why there is a sequential appearance of progressively more complex forms spanning many millions of years, and why there are transitional fossils — that is, fossils that are transitional between one major group and another. We can also see why they argue so vehemently that the earth is relatively young; who needs millions of years to produce the geological column if they can evoke a good old catastrophe? Only evolutionists need lots of time because they say that the evolutionary process is very slow.

The creationist's next problem is to find a suitable catastrophe to account for the rapid burial of all the creatures that are now found as fossils. The answer, of course, is found in the book of Genesis, in the Great Flood of Noah. This is not the first time that the Noachian flood has been woven into the fabric of a theory, but I doubt whether it has ever been done with such vivid imagination as Dr. Morris's account. According to him, so violent was the flood, and the upheavals that accompanied it, that every living thing was destroyed, save those riding out the tempest in Noah's ark. The bodies of the victims sank to the bottom of the waters, becoming enclosed in sediments and thus fossilized. The sequence in which they occur in these sediments (which hardened into rocks) is merely a function of where they were when they drowned, or of mechanical sorting. Dr. Morris tells us, for example, that mammals and birds are found highest in the geological column partly because they lived at higher altitudes than the lower tetrapods (amphibians and reptiles), and partly because they were better equipped to flee before the rising floodwaters. The reason the simpler organisms are found at the bottom of the column is that, being spherical, they tended to settle out first. When Dr. Morris tells us that "there is no observational fact imaginable which cannot, *one way or another*, be made to fit the creation model," (my italics) he is certainly not kidding!

To summarize the essentials of the creationist thesis:

- All living things were created at the same time, and therefore lived contemporaneously. No new organisms appeared after the creation.
- Earth history is dominated by catastrophism. The Flood of Noah, which covered the entire globe with water, extinguished all life forms that were not saved by the ark.
- Fossils are the remains of organisms that perished in the flood. Their arrangement in the geological column has nothing to do with time (they were deposited at essentially the same instant) but is due to the altitude they were living at prior to death, and also to mechanical sorting.
- The earth is young, probably about ten thousand years old, certainly not millions of years old.

Evolutionists disagree with each of these points. As we have said before, the process of evolution is too slow to be observed during our short tenure on the earth, and it follows that our evidence for evolution is largely historical. This book will therefore be largely, though not exclusively, concerned with fossils and the rocks in which they are found.

From the theory of evolution the following predictions about the fossil record can be made and tested:

- The earliest fossils are the simplest ones.
- There is a sequential appearance of various major groups of organisms throughout time.
- Intermediate fossils linking major groups will be found.
- The earth is old (probably about 4.5 billion years). One more prediction can be made which does not directly arise from the theory of evolution:
- While some fossils doubtlessly were formed catastrophically, as when buried by landslips, or washed away by floods and the like, most were probably formed under steady-state conditions similar to those pertaining today, as were the sediments that enclose them.

Before playing the creationists' game and testing the validity of the two "models" (evolution and creation), we need to explain just what evolution is all about. I can think of no better way of doing this than to trace the steps that led Darwin to formulate his theory of evolution by natural selection, as we shall do in the next chapter.

TWO   *Why Evolution?*

WHEN DARWIN SET OFF on his voyage aboard the *Beagle* in 1831, he had no preconceived ideas about evolution. True, he had heard of the evolutionary notions of Lamarck — that animals passed on to their offspring features they had acquired during their lifetimes (the giraffes, for example, got their long necks over generations of stretching up into trees for food) — and he had no doubt read the evolutionary speculations of his own grandfather, Erasmus Darwin, but none of this had made any lasting impression on him. Like almost everyone else of his time he believed in the permanence of species, and had no reason to doubt that they had all been brought into existence by the hand of God. In later years he recalled: "Whilst on board the *Beagle* I was quite orthodox [in religion], and I remember being heartily laughed at by several of the officers (though themselves orthodox) for quoting the Bible as an unanswerable authority on some point of morality." But Darwin's belief in the infallibility of the Old Testament had begun to change soon after his return to England: "I had gradually come by this time, i.e. 1836–1839, to see that the Old Testament was no more to be trusted than the sacred books of the Hindoos."

What could bring about such a radical change in a man's thinking? What did Darwin see during the voyage which caused him to deny the fixity of species, and the Genesis account of the Creation, in favor of evolution? He was a meticulous observer, and he saw more of the living world during his five-year circumnavigation of the globe than most of us see in a lifetime. Observation is an essential part of science, and the ability to be a careful and critical observer is a skill that is acquired through practice and training. So many people look without ever seeing, a shortcoming so frequently displayed by creationists.

We have used the term species, and most of us have the same understanding of the term — we recognize that robins, starlings, and cardinals are different species of birds — but we now have to define the term more precisely. A species may be defined as a group of similar organisms which freely interbreed with one another when they come into contact. This is the biological species concept. The sparrows in my garden are not likely to come into contact with Montreal sparrows, far less with sparrows from San Francisco, but if they did they would be able to freely interbreed and produce offspring because they are all the same species. Our definition recognizes the fact that the individuals of a species do not necessarily all live in contact — they usually live in small populations which are scattered throughout the range of the species (the species range is the geographical area where that species is found). My sparrows may belong to a population of several dozens of individuals, and there are probably many such populations living in North Toronto. The yardstick of a species, then, is whether interbreeding freely takes place. Interbreeding sometimes takes place between separate species, as between mallard and pintail ducks, but this is unusual and typically occurs in captivity when the individuals are not free to exercise their choice.

In the early days of classification there was no biological species concept; consequently mistakes were made. Linnaeus himself, the father of classification, who introduced the principle of giving organisms a double-barreled scientific name (*Homo sapiens* — man; *Canis familiaris* — dog), placed the female mallard into a separate species from its gaudy mate. This is a reasonable mistake for somebody who had not seen the two in the natural state.

Some species are much more similar to one another than they are to other species — lions and tigers are more similar to each other than either is to a fox or wolf or a bear, and are therefore placed into the same category of classification. The lion and the tiger, accordingly, are placed in the same genus, *Panthera* (*Panthera leo* — lion; *Panthera tigris* — tiger), which in turn is united with other cats and placed in the family Felidae. Foxes and dogs are similarly placed in the same family, the Canidae, and these, together with the cats, bears, badgers, seals, and many other meat-eaters, are placed in the order Carnivora. Closely similar species are interpreted as being closely related, evolved from a common ancestor. While closely related species can usually be distinguished from one another, this is not always so, and such exceptions are called sibling species. There are, for example, several species of birds called flycatchers belonging to the genus *Empidonax* which we cannot tell apart visually, but the birds know the difference, and that is the important thing! They recognize each other by a difference in song. A female of one species is disinterested in the song of a male of a sibling species; therefore, interbreeding is avoided and the species keep themselves quite separate.

The *Beagle* spent much time in South American waters, and while Captain FitzRoy busied himself with his surveys of the coast, Darwin explored ashore. The young naturalist was overawed by his first glimpse of a rain forest, with its teeming denizens and their struggle for survival. Each new excursion and new landfall assailed his senses with something unfamiliar and exciting. He filled his journal with notes, and his small cabin with specimens. Most of the animals and plants he collected were new to science, and these were periodically shipped off to England to be studied by various specialists. The change in fauna (animals) and flora (plants) from one region to the next was often quite remarkable. On an overland excursion from Buenos Aires to Santa Fé, Argentina, Darwin wrote:

> I was surprised to observe how great a change of climate a difference of only three degrees of latitude . . . had caused. This was evident from the dress and complexions of the men . . . the number of new cacti and other plants — and especially from the birds. In the course of an hour I remarked half-a-dozen of the latter, which I had never seen at Buenos Aires. Considering that there is no natural boundary between the two places and that the character of the country is really similar, the difference was much greater than I should have expected.

Why should regions which are physically so similar have their own species of birds? Why should the Creator have been so bountiful with his creative works? We can visualize Darwin pondering over these questions each time he saw something unexpected.

Darwin was every bit as interested in geology as he was in natural history, and spent much of his time studying rocks and collecting fossils. Before sharing some of his geological experiences, we should understand how fossils were interpreted in the 1830s. That fossils were the remains of organisms that had once lived on earth had long since been recognized, at least by intellectuals, and, largely owing to the studies of the French scientist Georges Cuvier (1769–1832), it was generally believed that many of these organisms were now extinct. For some people, however, extinction was unacceptable on religious grounds. To suggest that some creations had failed to survive was to suggest that they were imperfect, casting doubt on the perfection of the Creation and of the Creator.

There was, however, a way out of this dilemma. Cuvier attributed extinctions to changes in the environment — floods, cold weather, droughts, and the like. He believed that such catastrophes were both fairly frequent and localized. The Reverend William Buckland (1784–1856), England's first professor of geology, preferred to equate catastrophe with the Great Flood of Noah. He spent most of his life proselytizing this scriptural view. In later years he realized that he had been wrong, and that all the geological phenomena he had attributed to the Genesis flood were best explained in terms of the action of glaciers. How odd it is that we should be seeing another "scientific" revival of the Genesis flood, more than a century after Buckland's death.

While geologizing along the coast of Argentina, Darwin uncovered the remains of giant land mammals, including giant sloths and an animal similar to the llama. He recognized the close similarity between these gigantic mammals and their diminutive living relatives, and correctly surmised that they had become extinct in fairly recent times. The reason for the demise, however, was a complete mystery to him because he could find no evidence for any catastrophic changes in the environment: "Since their loss, no very great physical changes can have taken place in the nature of the country. What then has exterminated so many living creatures? . . . there are no signs of violence, but on the contrary, of the most quiet and scarcely sensible changes."

What Darwin saw clearly did not fit the scriptural account of animals perishing in the Great Flood.

Having spent more than three years voyaging back and forth along the coast of South America, the *Beagle* set sail for the Galapagos archipelago, 1100 km off the coast of Ecuador. Darwin's first impressions of these black volcanic islands, sweltering under an equatorial sun, were unfavorable. But they began to work their charm on him and the month he spent there was probably the most important part of the whole voyage. The archipelago, which is only about 200 km across at its widest point, comprises thirteen major and six minor islands, together with scores of nameless islets and rocky outcrops. Many of the islands are in sight of others, the distance between them often being less than 20 km.

Not surprisingly, the animals and birds of the Galapagos are reminiscent of those of the mainland. "It would be impossible," wrote Darwin," for any one accustomed to the birds of Chile and La Plata to be placed on these islands, and not to feel convinced that he was ... on American ground." But a large number of species are endemic to the archipelago — that is, they occur there and nowhere else. There are, for example, 228 endemic plant species, 28 birds, 32 reptiles, 4 rodents, more than 60 fishes, and more than 50 land snails. Some islands even have their own endemic species, the lava lizards being the most visible of these. These little lizards, which are about 20 cm long, are very conspicuous and can be seen on most islands, basking in the sun or scurrying across the lava. There are seven species; the six major islands each have their own species, while the seventh occurs on several of the major islands. Even more impressive than the lizards, at least during Darwin's time, were the tortoises, from which the islands got their name. Unfortunately these have been all but exterminated, but there was a time when they roamed in the hundreds. Fourteen races or subspecies have been recognized, and these were so distinctively different from one another that the vice-governor of the islands assured Darwin that he could tell from which island a tortoise came by just looking at it.

Why should so many of the species be unique to the Galapagos archipelago? Was it simply that the Creator had chosen to perform so many separate acts of creation, or was it because the islands had been colonized by species from the mainland and that these colonists had undergone modifications over the thousands of years since the islands formed?

We may imagine the young Darwin sitting on a cinder cone turning the question over in his mind. Marine iguanas bask in the sun between feeding forays in the sea (no other reptile in the world feeds on seaweed). Cactus plants, the first vegetation to get a foothold on the emergent land, bristle from cracks in the lava. It is not difficult to visualize a time when the lava was still molten and sulphurous. Two conflicting ideas jostle in his mind: the fixity of species and the modification of species. When he began his odyssey four years before, he had been convinced of the permanence of species, and the scriptural account of the origin had seemed quite satisfactory — but now? He had been deeply impressed by four observations: the way in which closely allied South American animals replaced one another as latitude increased; the discovery of large fossil mammals in South America that closely resembled smaller mammals living there now; the South American character of the Galapagos fauna and flora; and the way that so many of the islands had their own endemic species. These facts could only be explained by recognizing that species had become modified, and this idea in turn pointed to evolution. His speculations probably went no further, but, soon after his return to England the following year, he began collecting every scrap of information that might have a bearing on the question of how new species might evolve.

The key issue was the modification of species. He rejected the Lamarckian theory of modification by the inheritance of acquired characteristics, and therefore had to seek an alternative mechanism. He was aware of the powerful effect of selective breeding — one has only to think of the enormous variety of domestic dogs to realize its potential — and he wondered how such a force might operate in nature. The solution came to him after reading a book on populations. Organisms produced far more offspring than could survive — he had seen the struggle for existence so many times during his voyage — and if some organisms possessed favorable features, they would tend to be preserved over the less favorable variants. We all know that offspring are similar to, but not identical with, their parents, and it follows that some individuals have features that give them an advantage over their fellows. These advantages, however slight, give the individual a better chance of surviving. Darwin coined the term "natural selection" to describe this process of the differential survival of individuals.

The basic principles of the theory of evolution by natural selection are that individuals vary (a flock of sheep might look the same to us but not to an observant shepherd), that some of these variations increase an individual's chance of survival and hence of leaving more offspring, and since their offspring inherit their parents' features, they too have an increased chance of survival. The action of natural selection, operating over a long period, would cause a species to become better adapted to its environment, and hence to change. Darwin believed that the environment was slowly changing — he had seen evidence of change in the geological record — and as a species modified to this change it eventually became a new species; the alternative was extinction. He therefore expected to see finely graded sequences in the fossil record, sequences of old species undergoing gradual modifications into new species. However, when he turned to the fossil record for confirmation of his theory, he found not a single example of these sequences: "Geology assuredly does not reveal any such finely graduated organic chain; and this, perhaps, is the most obvious and gravest objection which can be urged against my theory." Darwin attributed his dilemma largely to the imperfections of the fossil record (he devoted a whole chapter to the subject), and many evolutionists since then have made the same point. The creationists, however, argue that this is just an excuse, and document their case with quotations from various paleontologists attesting to the richness of the fossil record. Was Darwin wrong?

Before answering this question, two points must be understood. First, most evolutionists now believe that species are essentially stable over long periods of time and that new species arise from existing species in a relatively short interval of time. Second, the fossil record lacks the fine resolution needed to show minor evolutionary changes. Darwin was therefore wrong in trying to use the fossil record to document the origin of new species, but he was not wrong in choosing evolution over special creation. Had he used the fossil record to document major steps in evolution — for example, the fact that the earliest fossils are the least complex — a purpose for which it is ideally suited, he would have found what he was looking for. The next two chapters examine these critical points in more detail.

THREE   # *Evolution: Past, Present, and According to the Creationists*

 DARWIN'S CONCEPT OF EVOLUTION, as we saw in the last chapter, was one of continuous change. He believed that the environment was continually changing and that species were continually responding to this change by becoming better adapted. Therefore, if the history of a particular species was monitored over a period of time, it should be seen to be undergoing continuous changes. These slight changes would eventually accumulate to the point where the current species was sufficiently different from the original species to have become a new species. Darwin wrote: "Natural selection acts, as we have seen, exclusively by the preservation and accumulation of variations, which are beneficial under the organic and inorganic conditions of life to which each creature is at each successive period exposed." The alternative fate for a species was for it to fail to adapt, and therefore to become extinct. Darwin recognized that extinction was an integral part of the evolutionary process; there have to be losers as well as winners in the struggle to survive.

While the gradual transformation of one species into another over long periods of time does seem to occur (fossil sequences which support this have now been found), this is probably the exception rather than the rule. Most species, instead, appear to be essentially stable over long periods of time. That is not to say that species do not respond to minor fluctuations in the environment, because they do, but over the long term — hundreds of thousands or even millions of years — they remain unchanged. How do we arrive at such a position when the very word evolution implies change? This is one of the questions we will be answering in this chapter.

We have said in the previous chapter that a species is a group of organisms that freely interbreed, and therefore freely exchange genetic material among themselves. At the same time they maintain themselves separate from other species by not interbreeding with them. However, individuals within a species, as we have seen, are not identical to one another; some are better able to survive than others, and we call these the fittest. This year's criteria for fitness may not be the same as next year's, because environmental conditions fluctuate. Let us illustrate this point by reference to winter mortality in warm-blooded animals.

Birds and mammals maintain high and constant body temperatures by virtue of their high rate of body chemistry (metabolism). Although there are all manner of advantages in this, it is a very expensive business, because high metabolic rates require the consumption of large quantities of food. Birds and mammals have to eat the same amount of food on winter days as they do on summer ones, perhaps even more, and this poses grave problems for resident birds and mammals during harsh winters. Heat is lost from the body surface, and it is therefore advantageous to be a large individual because large bodies have a relatively smaller surface area for their volume. This is one of the reasons why we have to take such care to keep babies wrapped up. However, if an individual is too large, he begins to lose his advantage simply because large individuals eat more food than small ones. Natural selection therefore works against small animals, and also against very large ones; if we surveyed animal populations before and after a hard winter we would find that mortality was highest in small individuals, and also high in the largest individuals. The result of all this is that there would be a shift in all populations towards a large average body size. If, for example, the average body length of some particular species of mammal was 7.5 cm in one summer, it might be 8.2 cm the next summer following a harsh winter.

This trend would reverse itself in mild winters; consequently, if we monitored the body sizes of small mammals and birds over many years we would see some ups and downs, but there would not be a continuous trend in any one direction. The response of populations to fluctuating environmental conditions is brought about by the action of natural selection, but, because these trends are short-lived, and because the individuals of a given species continue to interbreed, the trends do not have the potential to lead to the evolution of a new species. Mice get bigger and mice get smaller, but they are still the same species of mice.

Our example of fluctuating body size deals with only one variable, but there are many more that could be measured. A series of studies on wild populations of house mice in Britain, which extended over a period of sixteen years, showed that individuals differed from one another in far more ways than just their skeletal features. These differences included social behavior, reproductive condition, blood chemistry, metabolic rate, the weights of various organs including the heart, kidney, spleen, adrenal gland, thymus gland, and fat reserves, and even how well individuals insulated their burrows. One particularly interesting result of the studies was that features which were advantageous to individuals one year were not necessarily advantageous in other years. For example, on the Welsh island of Skokholm, mice with higher metabolic rates had an increased chance of surviving the winter of 1967, but this did not hold for the following year. This graphically illustrates the point made earlier that features which are advantageous at one time are not necessarily advantageous at another; evolution is opportunistic.

Many species, our own included, have an enormous species range, while others, like the seven species of Galapagos lava lizards, have very small ones. Animals and plants occupying large species ranges obviously encounter a much wider spectrum of conditions than those in small ones. Imagine the enormous range of environmental conditions that our own species encounters — from polar wastes to tropical rain forests. The individuals of a wide-ranging species, under the action of natural selection, become adapted to local conditions, and this results in what we call geographic variation. Caribou that live in the tundra, for example, have relatively larger hoofs than those living in the boreal forest to the south; they have a white winter coat and are somewhat smaller. The people of the tundra, in contrast to those living in the tropics, are stockily built and have fair skin — but more about skin color later.

So far two points have been made. First, over a period of time a species undergoes change in response to fluctuating environmental conditions, but these changes are not directional; rather, they fluctuate about the mean. There is, therefore, no permanent change in the species. Secondly, we have seen that local populations of species with large ranges become adapted to local conditions, and this phenomenon, which has been recognized since long before Darwin's time, is called geographic variation. Neither environmental fluctuations nor geographic variation lead to the origin of new species. If we could look at all the individuals of a given species that have lived in a particular area for the last million years, say all the beavers from northern Michigan, we would probably not be able to detect any significant differences between them. Indeed, a study of European mammals has shown that more than half of the living species can be traced back for some half-million years without any obvious changes having taken place, and about seven per cent can be traced back without change for about three million years.

There are also many examples of genera, the next highest category after species, that have remained essentially unchanged for millions of years. These representatives of ancient groups are generally called "living fossils." Modern brachiopods, or lamp shells, which look something like clams, hardly differ from the brachiopods that lived in the Cambrian period, 500 million years ago. The brachiopod genus *Lingula*, for example, found burrowing in the mud along tropical shores, is actually found in Cambrian rocks and has not undergone any change (certainly not in its hard parts) in all that intervening time. The reptile *Sphenodon*, which looks like a lizard and is today restricted to a few islands off the coast of New Zealand, is closely similar to genera that lived during the age of dinosaurs; it has survived 200 million years without any significant change. Many other examples of living fossils can be given: the horseshoe crab *Limulus*, unchanged since the Triassic (220 million years ago); the Port Jackson shark *Heterodontus*, and the Ginkgo tree *Ginkgo*, unchanged since the Jurassic (120 million years ago); the crocodile *Crocodilus* and the Australian lung-fish *Epiceratodus*, unchanged since the Cretaceous (65 million years ago); the marine turtle *Chelonia* and the lizard *Lacerta*, unchanged since the Eocene (60 million years ago).

If we broaden our category to include major groups of organisms, orders, and classes in the classification system, we find that stasis, that is, no change, is the rule rather than the exception. There appear to be no important differences between modern representatives of major groups and their ancient relatives. To give just a few examples, modern jellyfish are similar to their Carboniferous (300 million years ago) relatives, turtles have changed little from the Triassic period (200 million years ago), the coelacanth fish is very similar to its relatives from the age of dinosaurs (200-65 million years ago), sharks have changed little since the Jurassic (180 million years ago), modern bony fishes are very similar to Cretaceous (65 million years ago) ones, and insectivorous mammals, such as shrews, are very similar to their Eocene (60 million years ago) relatives. I can almost see the creationists nodding in agreement and saying that this is exactly what their model predicts, but I hasten to point out these various groups we have mentioned did not appear all at once (see Chapter 9).

Stasis is a demonstrable fact, and while this might appear to fly in the face of evolution, a little thought will show that it makes perfectly good sense. Let us illustrate the point by reference to the reptiles. Reptiles are similar to amphibians in their skeletal anatomy, and when we compare early reptile fossils with their amphibian counterparts we sometimes have difficulty distinguishing between them. The great evolutionary advantage that reptiles have over amphibians, though, is that they are fully terrestrial, and do not have to return to water to breed as almost all amphibians have to. Several features have contributed to this emancipation from water, including a dry skin that is impermeable to water, a good pair of lungs, internal fertilization, and an egg, called an amniotic egg, which provides the developing embryo with its own water supply. Once these basic reptilian features had evolved, they were retained and passed on without major change for tens of millions of years. We can also see many examples of stasis within the anatomies of the different types of reptiles which evolved from those early beginnings. Turtles, for example, are distinguished from all other reptiles in having their body encased in a bony shell, and once this structure had evolved, some 200 million years ago, it was passed on, essentially unchanged, to the present day. The modern concept of evolution, then, is one of rapid bursts of change followed by long periods of stasis. We can illustrate the logic of this with some analogies from the world of technology. The present-day internal combustion engine has not undergone significant change since its invention over a century ago, and the same is true of television sets, electric light bulbs, and sewing machines. Where significant changes have occurred, as in the development of modern aircraft and calculators, these can usually be attributed to the appearance of a new structure. In the two examples chosen, these major jumps can be attributed to the jet engine and to the micro-chip respectively.

To return to where we began this chapter, we have to ask ourselves how new species have originated if not by the accumulation of small changes over long periods of time. The key to the problem seems to lie in the question of interbreeding. We have already said that all the various species of organisms maintain their own individual identities because they do not interbreed. The other side of this coin is that the individuals of a given species are prevented from diverging very far from all the other individuals simply because they interbreed with one another. Darwin made this point in the *Origin of Species*, noting that when a farmer wants to perfect a particular variety of cattle he makes sure that his breeding stock does not interbreed with other varieties. Most evolutionists are agreed that the only way a new species can evolve is if a small group of individuals become separated from the rest of the individuals of that species. The most obvious place that this is likely to occur is at the edge of the species range, and for two reasons. First, peripheral populations are the most likely to become separated from all the rest. Secondly, organisms living at the very edge of the species range are likely to be the most divergent in terms of their adaption to local conditions. This is because the edge of the range marks the transition from an area where environmental conditions are correct for the species to areas where conditions are less suitable.

Do we have any examples of speciation occurring because of geographic isolation? We have already seen that the Galapagos Islands, separated from mainland Ecuador by 1100 km of ocean, have a high percentage of endemic plants and animals, and these are the modified descendants of mainland colonizers. Indeed, whenever island floras and faunas are compared with those of the closest mainland, endemism is found to be the rule rather than the exception. The prevention of interbreeding permits differences to accumulate in the two isolated groups until a point is reached where they are no longer able to interbreed even if they had the opportunity, and at this point a new species is said to have evolved. The parental species may well carry on just the way it was prior to the isolation event, but the isolated group has diverged. We have reason to believe that this period of divergence, the time required for the origin of a new species, is relatively short, thousands of years as opposed to millions. Do we have any evidence for this belief?

There is a small lake in Africa, Lake Nabugabo, which was once part of Lake Victoria but became separated from it by the formation of a sandbar. Based on geological evidence, the sandbar is known to be about 3,500 years old, so the new lake has not been in existence for more than this period. When the new lake was formed it was obviously stocked with fish from Lake Victoria, and most of the species found there today are identical to those found in Lake Victoria. There are, however, five species which are endemic to Lake Nabugabo and which have evolved from Lake Victoria species some time during the period of isolation. These species cannot therefore have evolved more than about 3,500 years ago. This may be a very long period of time in terms of our life experiences, but in geological terms it is a mere instant.

Stasis should not be considered as the antithesis of evolution. The fact that a given species of organism is surviving is evidence that it is well adapted to its habitat, and if no major environmental change occurs the species will continue just the way it is. The occasional harsh winter is obviously not a major environmental change, but the onset of a new ice age, say, would be, and this could well lead to the origin of new species. In the second half of the Pleistocene ice age, for example, a new species of elephant evolved, the mammoth, which differed from modern elephants in several features, including the possession of a thick, shaggy coat.

To summarize the view of evolution that most evolutionists share today:

- Species remain unchanged (except for minor fluctuations) over long periods of time, for tens of thousands of years and perhaps for several millions of years.
- New species probably evolve only when a segment of the population becomes isolated from the rest.
- Speciation occurs relatively rapidly, probably in a matter of only a few thousand years and possibly less.

This concept of long periods of stasis interspersed with periods of rapid change is often called "punctuated equilibrium." The old concept of continuous change, as envisioned by Darwin, is often called "phyletic gradualism." We should notice that in the old view one species gradually changes into a new species. In the modern view, however, a new species rapidly diverges from the parent species, which then carries on as before. We should note that supporters of the concept of punctuated equilibrium do not deny that species sometimes evolve by phyletic gradualism; they just consider that this is the exception rather than the rule.

As we might expect, the creationists' views of the origins of new organisms is quite different. When Dr. Gish tells us that nobody has ever observed "the conversion of a fish into an amphibian, or an ape into a man," I conclude that his concept of evolution is one of continual change — the slow transformation of one type of organism into another — and this suggests that he rejects current concepts of evolution. At this point let us say a few words about the notion of man having evolved from apes — a popular misconception. Few evolutionists would suggest that we descended from apes — what evolutionists would say is that we share a common ancestor with apes. There is a world of difference between the two statements. Consider the following analogy:

"I am descended from my older brother."

"My older brother and I are descended from a common ancestor (we call him father)."

Dr. Gish's use of phrases like "amoeba-to-man" and "fish-to-Gish" to describe the theory of organic evolution is also revealing. Such sentiments, suggesting, as they do, that there was some striving on behalf of primitive organisms to give rise to advanced ones (these ideas are described as teleological), and ultimately to man, are not in accord with an evolutionist's understanding of evolution. Supposing I could trace my ancestry back to Oliver Cromwell. Would it make more sense to say that Oliver Cromwell's major objective was to give rise to me, or that I just happened to be one of the probably thousands of people who is descended from him?

To return to the "amoeba-to-man" notion: some people might like to think that man stands at the very pinnacle of evolution, and that our appearance on earth was what evolution has all been about, but that is not the way I see it. As far as I can see, man is one of well over one million species that are living on the planet, one twig on a very dense bush rather than the angel at the top of a Christmas tree. That is not to deny man's dominant and unique position in the world, but merely to put him into larger perspective. I think that our behavior towards one another, and towards the rest of the living world, is more readily understood if we think of ourselves as clever primates with an aggressive streak, rather than as superior beings standing at the very summit of evolution.

An interesting feature of the creationists' view of evolution is that they accept that variation directed by natural selection can lead to minor change. They insist, though, that this mechanism can never produce anything new. As far as they are concerned, variation and natural selection may have brought about the modification of a mainland Ecuadorian lizard into a Galapagos lava lizard, but the end result is still a lizard. This is true, of course, but creationists overlook the tremendous amount of variability that exists within a species, and they underestimate the powerful force of selection. Individuals of a given species may look all alike to us, but this is only because we are not looking at them very carefully. Anyone who has sat down and made a detailed examination of a large number of individuals of one species, be they butterflies, snails, birds, plants, or whatever, knows that a lot of variation exists. Aside from the obvious variability in external features such as size, color, color patterns, and body proportions, there is considerable variability in the internal structures. Bones vary in surface details and relative proportions, organs vary in size and relative position, blood vessels vary in the details of their branching patterns, and much more besides. But these are only anatomical features. If we studied the living organism, we should be able to see differences in behavior, feeding success, running speed, not to mention differences in the numerous internal body processes, such as digestion, respiration, temperature regulation, and the like, processes which are collectively described as the organism's physiology. The more closely we look, the more variability we see.

Darwin recognized the importance of variability, and he knew what tremendous potential this variability gave to the process of selection, whether natural or artificial. He chose to illustrate his point with artificial selection, that is, by selective breeding rather than natural selection, simply because the former is more rapid than the latter. Consider the enormous variety we see in dogs, from the tiny Chihuahua to the hulking St. Bernard, or the incredible variety of pigeons. The differences between the different breeds of pigeons are greater than the differences between most species of wild birds, and the same is true for most other domestic species (pigs, sheep, cattle, chickens, and the like). The only reason the various breeds within these domestic species are all referred to as the same species is because we direct their breeding.

Dr. Morris goes so far as to say that the modification of species through natural selection, rather than explaining evolution, is a marvelous example of the Creator's way of enabling the species that He created to survive in nature! He goes on to argue that even if variation and natural selection could produce a new structure, like a wing, this would be useless or even harmful until it was fully developed. We shall see in Chapter 10 just how groundless this argument is.

Though creationists have some grasp of the concept of variation and natural selection, it is a tenuous grasp, as the following example shows. According to Dr. Morris, one of the most vexing questions to modern evolutionists is the origin of the races of man. If man evolved from a common ancestor, and if "no one race is better than another, as most modern evolutionists affirm, then how did they ever get to be so different in appearance?" He suggests that intellect and "physical capacities" (whatever that means) should surely have had a greater survival value than "such relatively innocuous differences as skin coloration." He goes on to point out that thoughts such as these lead to racism, and that modern evolutionists "quite rightly repudiate racism . . . even though this leaves them with an unsolved scientific puzzle."

Before we set about solving the puzzle (though such a simple problem hardly merits description as a puzzle), we should point out that Dr. Morris's reference to racism follows on from several quotes from evolutionists, including T. H. Huxley. Huxley, a contemporary of Darwin, was an active campaigner for evolution, and in some of his writings he refers to the superiority of the white races. Such remarks are indefensible, but we should remember that our nineteenth-century forefathers were less sensitive to commentaries about racial differences than we are today. Dr. Morris is very quick to point out that modern evolutionists repudiate racism, but this is only done after a connection has been made between evolutionists and racists. I am reminded of the strategy of the wasp, which removes the sting quickly but leaves the poison in the wound.

As I see it, Dr. Morris has set a trap for the evolutionist: either show me that such "relatively innocuous differences" as skin color are more significant than intellect, or show yourself to be a racist. We will now deal with both problems in their turn, the evolution of skin color and the evolution of intellect.

A couple of years ago a colleague and I visited the Galapagos archipelago. It is important for the story to note that we are both Anglo-Saxons, but that he tans much faster than I do. The Galapagos Islands are strung right across the equator, and I could hardly believe how hot the sun was. I kept well covered for the first week, but I still managed to get sunburnt shoulders, even though I had resorted to wearing two shirts. My wrists and the backs of my hands suffered most, and I had several painful blisters which only cleared up during the second week. Meanwhile my colleague became a magnificent shade of brown, though he too was unable to stay in the sun without a shirt for any length of time. Our Ecuadorian crew, with the beautiful golden-brown skin with which they were born, could stay shirtless beneath the equatorial sun with no discomfort. There is no question who had the greatest advantage in terms of skin color. The Ecuadorians could have stayed out in the sun all day without any ill effect. My colleague would have got sore had he stayed out all day during the first week, and if I had tried it I might well have died of heat stroke. Now to get back to Dr. Morris and the "unsolved scientific puzzle." Surely it is not too difficult to correlate races with geography. White men are found naturally in the colder climates, black men where it is hot, with all shades of brown in between. We are a geographically variable species because our species range covers the whole world.

What about the problem of the evolution of the intellect? As an evolutionist, I suppose I am meant to believe that highly civilized races have a greater intellect than "primitive" races. Actually this old chestnut has cropped up before. Alfred Wallace, who independently arrived at a theory of evolution like Darwin's, could accept that most human features had evolved through the mechanism of natural selection, but could not accept this for the evolution of the human brain. He recognized that the intellectual capacity of a "primitive" tribesman was the same as his own, but could not see how this could have arisen through natural selection. The tribesman, with his simple way of life, is not receiving any advantage from his untapped intellectual powers, so how could his intellect possibly have evolved? The solution to the problem is not so difficult to work out. The tribesman *is*, of course, enjoying a selective advantage from his intellect. His brain allows him to solve everyday problems, like stalking prey, gathering food, and planting crops, and the fact that his brain is capable of much more besides is irrelevant. A radio set is capable of receiving more than one station, but we are still deriving benefit from it when we only listen to one program.

Not only is Dr. Morris puzzled by the evolution of skin coloration, he also has problems in reconciling "living fossils," like crocodiles and ginkgo trees, with evolution. He makes the point that there has been so little change in such organisms that it is hard to believe that the theory of evolution is valid.

As we saw at the beginning of this chapter, though, stasis is in perfect accordance with the modern concept of evolution: without environmental change there is no evolutionary change. Creationists should familiarize themselves with post-Darwinian literature before writing critiques of evolution.

We have seen that the basic mechanism of evolution is that there is variability among individuals, and that this variability, under the direction of natural selection, can lead to the origin of a new species. An obvious question to ask now is how individual variability arises, and whether it is sufficient to account for the appearance of major evolutionary changes over long periods of time.

Most of the characteristics of an individual, from the color of its eyes to how efficiently it can digest food, are determined by its genes. While some organisms reproduce asexually, as when an amoeba divides into two identical daughter cells, most organisms reproduce sexually, and this is a large source of variability. To see why this is so we need to say a little more about genetics. Genes can be thought of as sets of blueprints, arranged in patches along the length of a slender thread called a chromosome. The chromosomes lie within the nucleus of the cell (though not all cells have a nucleus) and this is essentially a capsule which is separated from the rest of the cell by a thin membrane. Every nucleated cell in our body, which is almost all of them (mature blood cells do not have a nucleus), has 46 chromosomes. If we examined our chromosomes under a microscope, we would see that they varied in length. Careful study reveals that there are two of each kind; we therefore have 23 pairs of chromosomes, hence two sets of all the various genes (to be precise, females have 23 similar pairs but males have only 22 because the 23rd pair, the XY sex chromosomes, are not similar). During the formation of gametes — eggs and sperms — the cells undergo a reduction division so that each gamete has only one set of chromosomes. This all makes good sense because when a sperm, with 23 chromosomes, fertilizes an ovum, also with 23 chromosomes, the resulting cell, which will eventually become a new individual, has a complete set of 46 chromosomes. Imagine that the husband's chromosomes are all colored blue and that his wife's chromosomes are all red. Their offspring would have one set of 23 red chromosomes and one set of 23 blue chromosomes.

Now imagine the baby grown up, and producing eggs of her own. When her cells, each with 23 red and 23 blue chromosomes, all jumbled up of course, undergo the reduction divi-

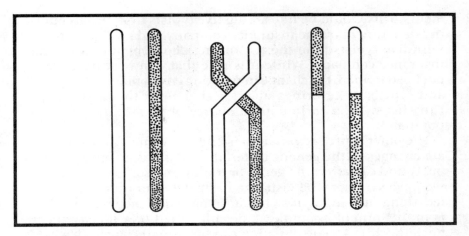

Crossing over between chromosome pairs.

sions to give egg cells (each with only 23 chromosomes), think of all the possible combinations. One egg may have 23 blue chromosomes, another may have 22 blue and one red, another 21 blue and two red and so forth. Remembering that each of the 23 chromosomes is different from every other one, you can see that the number of possible combinations is enormous. Her husband similarly produces millions of unique cells (sperms), and any one of these will fertilize one of her eggs. It is hardly surprising that, aside from identical twins (formed when a single fertilized egg separates into two cells, each one of which develops into a fetus), no two people in the whole world are identical. This shuffling up of chromosomes prior to fertilization is one source of variability, but there are two others.

First, there is a process called crossing over, which is a common occurrence during the formation of the gametes. During the reduction division the chromosomes line up in their pairs, and quite often they break in the same region. The two separate segments of each chromosome then join up with the corresponding segment of the other member of the pair. Imagine two sausages lying side by side, a red one and a blue one. A knife descends, lopping 2 cm off the right end of each sausage. The end pieces are switched and glued back in place. We now have a red sausage with a blue end piece and a blue sausage with a red end piece. The result of crossing over is that the individual genes which were once part of one chromosome become part of another chromosome (though this is one of the same pair), and this causes a change in their genetic environments. This would be like taking five typists from the accounting department of one insurance company and swapping them with five typists from the accounting department of a second insurance company. While it is true that a new structure has not been formed, the change in working environments for genes and typists alike brings about a change in their operation. Crossing over is an important source of variability within a population.

The other source of variation is effected by mutations, which are changes in the genetic material. Mutations arise spontaneously and can affect the genes or the chromosomes. The genes, we have said, may be visualized as patches of blueprints dotted along the length of a chromosome. The blueprint material is in the form of a complex molecule called DNA (deoxyribose nucleic acid). A gene mutation can be a small change in a part of one of the DNA molecules, in which case it may have little effect on the individual, though this does not necessarily follow. Alternatively the mutation may be a large change, in which case the individual will be markedly affected. Chromosome mutations include pieces being lost, pieces being broken off and attached to an entirely different chromosome, and chromosomes being duplicated. Down's syndrome in our own species is usually caused by the spontaneous duplication of one of the chromosomes (No. 21). Children affected by Down's syndrome have 47 instead of 46 chromosomes. There are several other chromosomal abnormalities, usually arising through the spontaneous duplication of chromosomes, and all are major in their effect. There are also numerous genetic mutations, with a wide range of effects. The molecular structure of human hemoglobin (the pigment that gives blood its red color), for example, has been found to have more than one hundred variants and each one of these is caused by a particular gene mutation. While some of the variants appear to have no obvious effects at all, others cause various forms of ill health.

The harmful effects of mutations is a point on which creationists focus a great deal of attention. How can evolution possibly work, they ask, if mutations, which are the source of new genetic variability, are almost always harmful? This does seem like a formidable problem for evolutionists, but there are two important points which creationists overlook. First, geneticists who study evolution are agreed that by far the greatest source of genetic variability within sexually reproduced organisms is due to crossing over and to the shuffling up of chromosomes that precedes the formation of gametes, rather than to mutations. Secondly, a given mutation cannot be assessed as being harmful or beneficial unless the environment is also taken into account. A mutation that is harmful to an organism living in its normal environment may be beneficial if the organism is moved to a new environment, or if its own environment is changed. Not only does the external environment have to be taken into account, but the internal or genetic environment of the mutated gene also has to be considered. A mutation arising among one particular suite of genes may be harmful, but in another group of genes it may be beneficial. This is because the genetic environment, as we have already said, has an important influence on the operation of individual genes.

The arguments that a mutation may be harmful in one environment but beneficial in another may not sound very convincing, but we have plenty of evidence to support it. Since this is such an important point, we will spend some time looking at several of these pieces of evidence.

Man began seriously polluting the environment during the early part of the eighteenth century, at the start of the Industrial Revolution. One of the many manifestations of this damage was the disappearance of a wide variety of organisms. Salmon no longer returned to dirty rivers to spawn, birds kept clear of sooted skies, and flowers failed to bloom in contaminated soil. Some organisms did adapt to the change, however, through the mechanism of natural selection working upon genetic variability, and some of this variability is attributable to mutations.

The soils near lead, copper, and zinc mines contain these metals in high enough concentrations to kill plants, but some species, notably the grass *Agrostis tenuis*, which can tolerate lead and copper, and sweet vernal *(Anthoxanthum odoratum)*, which tolerates zinc contamination, are able to grow upon abandoned mine workings. The plants that grow on contaminated soils are in some way different from plants of the same species growing on normal soils, because if the latter are transplanted to contaminated soils they die. The differences between tolerant and non-tolerant individuals are certainly genetic, because seeds taken from tolerant plants, and raised in the laboratory, are tolerant of contamination, while seeds raised from non-tolerant plants are killed by contaminated soils. Natural selection favors plants which have the metal-tolerant gene in the contaminated soils, but works against such plants on normal soils. We know this because very few of the tolerant variety are found on uncontaminated soils. Under normal conditions, then, individuals which have the variant or mutant gene that gives metal-tolerance are at a disadvantage. However, when the environment is changed by the influx of metal contaminants, these same individuals find themselves at a considerable advantage.

The use of DDT spraying to control insect pests, which came into wide usage soon after the Second World War, was initially successful. This was because the vast majority of insects were very sensitive to this chemical. After a number of years, however, spraying became less and less effective. The insects were becoming immune to DDT. A similar build-up of immunity, to antibiotics, has been seen in bacteria, and this has created some serious health problems. The question which naturally arises is whether the pesticides or antibiotics have somehow changed the organisms, or whether the immunity was due to random mutations occurring within some of the individuals of the populations tested. Some elegant laboratory experiments have shown that the latter is the true situation, and we will now look at one of these experiments conducted upon bacteria.

Bacteria are grown in the laboratory inside shallow glass dishes called petri dishes. A nutrient jelly is poured onto the bottom of the dish and, when this has set, a solution containing bacteria is squirted or dripped onto the jelly. The lid of the dish is closed, to prevent contamination with air-borne bacteria, and the dish is placed in an incubator. After a few days the bottom of the dish is dotted by small patches, each comprising thousands of bacterial cells.

Each patch or colony of cells has been derived from the multiple divisions of only a few original bacterial cells, so each colony has only a few genetically different types, and these are a little different from the cells of other colonies. In the experiment on immunity, a piece of sterilized felt, which is exactly the same size as the bottom of the petri dish, is lightly pressed down onto the bacterial colonies. Hundreds of cells from each colony are thereby transferred to the felt, which is then carefully transferred to a second petri dish and pressed down gently onto the nutrient jelly at the bottom. By taking care not to rotate the felt disk during the transfer, the exact positions of the original colonies can be replicated in the second petri dish. The jelly at the bottom of the second dish is impregnated with the antibiotic streptomycin. Consequently, when the transferred bacteria are incubated for a couple of days, most of them die. Only those cells that are streptomycin-resistant survive, and these are the only ones that give rise to colonies of cells.

Instead of the dozens of thriving bacteria colonies seen before, there may be only one or two colonies in the second petri dish. Because these surviving colonies are in exactly the same relative positions in the second petri dish as the parent colonies in the first, these parent colonies can be identified. Did the thriving bacteria survive because the streptomycin induced some change in them, or did they survive because they inherited the streptomycin-resistant mutation from their parent cells? We predict that the second alternative is true and that cells taken from the parent colonies are also streptomycin-resistant. Cells are taken from the parent colonies that gave rise to the resistant colonies and cultured in a third petri dish with streptomycin-impregnated jelly. The bacteria flourish, confirming our prediction.

We will now look at a laboratory experiment that shows the beneficial effects of mutations in modified environments. This experiment is conducted upon the geneticist's favorite animal: the fruit fly (Drosophila). For an animal whose genetics have been so intensively studied, the fruit fly is a singularly unimpressive creature. You can see these 3 mm-long flies hovering around fruit bowls in the summer, and they owe their popularity among geneticists to a number of practical reasons. They are easy to keep, they breed rapidly, they have a large number of easily recognized mutations, and into the bargain, their larvae have nice big chromosomes which lend themselves to microscopic study. One of the mutations, called vestigial wing, causes the wings to be so stunted that they cannot be used for flying. This is obviously most detrimental in the wild, and individuals possessing this mutant gene fail to survive. If the environment is changed, though, the condition can become advantageous. In an experiment, small fly cages were set up such that when a fan was turned on outside the cages any flies that were on the wing were swept through the bars and thereby removed from their food source; they were selected against. Each cage contained a mixed population of normal and vestigial-winged flies and after a short time, as we might have expected, only vestigial-winged flies remained. While this experiment is rather simplistic, and could also be interpreted as fanciful, it is a fact that many flying insects which have colonized small oceanic islands have secondarily lost their wings. The advantage conferred by wings, and the ability to fly, is reversed when there is a high probability of being swept away from land on a sea breeze.

We have seen that the effect of a given mutation is dependent upon the external environment, but we now have to consider how its action is modified by the internal or genetic environment. Mutations do not necessarily have an effect on the bearer, because their action may be partially or completely swamped by the normal gene on the opposite chromosome (remember that chromosomes are paired and there are therefore two sets of each gene). There is, for example, a particularly harmful mutation in man that causes anemia, a condition where the blood does not carry out its functions properly. If both chromosomes bear the mutant gene, the individual has the condition called sickle-cell anemia. His red blood cells, instead of being round, are sickle-shaped, and the condition is usually fatal. If only one of the mutant genes is present, though, the red cells are normal but the individual has some symptoms of anemia and is therefore not as healthy as a normal person. However, in regions of Africa where malaria was once rampant, it was found that this mutation was present at an unusually high level. It was subsequently found that, in the single dose, the mutation gave immunity against malaria. Therefore, individuals who in non-malarial areas were at a disadvantage because of their mild symptoms of anemia found themselves at a definite advantage when they lived in malarial areas. This example illustrates how both the external and the internal environments can determine whether a mutation is harmful or beneficial. In non-malarial areas the mutant gene is always detrimental, but when the populations are exposed to the infection, the mutant is beneficial, provided that it is combined with the normal gene.

General statements to the effect that mutations are invariably harmful are clearly incorrect. Mutations are usually harmful to organisms exposed to their *normal* environmental conditions, but this situation can be reversed when environmental conditions change. This all makes very good sense when we stop to think about it. Any species that is alive and thriving is obviously well adapted to its environment, and has probably been in existence for numerous generations. All the possible mutations which can occur have probably already occurred many times over, and it is likely that all of the advantageous ones have been incorporated into the gene pool (all of the genes of all of the living individuals) of the species. Any mutation that arises, then, is almost certain to be harmful, provided, of course, that the environment remains unchanged. Before leaving the subject of genes, we should briefly discuss gene regulation.

There are basically two types of genes: structural genes, which direct the manufacture of the various components that comprise the body, and regulator genes, which control the structural genes. Regulator genes are responsible for switching structural genes on and off, and therefore determine the sequence and rate at which the developmental processes proceed during embryology. Changes in these sequences can have profound effects on the individual, and it is conceivable that major evolutionary changes have been brought about by relatively small changes in regulator genes. We will discuss gene regulation in Chapters 10 and 12.

As an evolutionist I do not profess to have answers to all the questions that have been raised by critics of evolution — far from it. There are many aspects of evolution that puzzle me as well as other evolutionists. There is, for example, a major discussion going on among evolutionists over the mechanisms of macroevolution, that is, the evolution of large-scale changes. Can major changes, like the appearance of lungs or feathers, be accounted for by the accumulation of lots of small changes, or is some other mechanism involved? Although there is frequent disagreement among specialists on the question of mechanisms, there is no disagreement on the question of whether evolution has occurred. The fact of evolution, as we shall see in Chapters 9 through 14, is clearly documented by the fossil record.

The creationists, of course, choose to ignore the facts, and stick doggedly to the strategy of trying to make evolution look impossible and creation inescapable. Consider their ploy of placing a single-celled organism down beside a man and challenging evolutionists to fill in the gap! Dr. Gish makes a cute joke that the instantaneous transition of a frog into a prince is labelled as a nursery tale, but the 300-million-year transition of frog to man, by evolution, is labelled as science. Ignoring the poetic license taken with relationships, we must ask what he achieves with these games. If we placed a 747 airliner down beside the Wright brothers' first glider, we would be hard-pressed to see an obvious connection, but would anyone doubt there was an evolutionary relationship between them?

How Did It All Start?
The Numbers Game

 WHATEVER ACCOLADES DR. CHANDRA WICK-
RAMASINGHE may achieve in his own field of
mathematics, I believe he will always be re-
membered for his pronouncement that life
could not have evolved because the odds
against it were just too high. When he took
the stand as the creationists' star witness at the recent Arkansas trial (a court action in which the American Civil Liberties
Union successfully proved that the state of Arkansas violated
the constitutional separation of church and state when it introduced a bill legislating the teaching of creation alongside
evolution), he put the odds against life having originated by
evolution at 1 to 10^{40000} (a one followed by 40,000 noughts). A
more graphic illustration of the odds was given to the English
press by his colleague, British astronomer Sir Frederick Hoyle,
who compared them with the chances of a gale blowing
through a junk yard and congregating all the pieces into a 747
airliner.

According to Dr. Wickramasinghe, the only logical explanation for how life formed was that it was created, and he treated
the Arkansas courtroom to a three-hour lecture on the subject.
Drawing on his joint researches with Sir Frederick Hoyle,
he outlined how the Creator had dispersed micro-organisms

throughout space. Some of these organisms had been carried to earth on the tails of comets, and so life had begun. The major evolutionary jumps that have been documented in the fossil record could then be explained by supposing that genes from outer space had showered down onto the earth and become incorporated into the genetic material of living organisms. After this discourse, Dr. Wickramasinghe was asked to read a passage from his recent book, *Evolution from Space*, co-authored by Sir Frederick. The passage in question contained the idea that insects may be more intelligent than humans but that they are concealing the fact from us!

Now when a mathematician and an eminent astronomer join forces to show that the origin of life from non-living materials is mathematically impossible, creationists sit up and take notice. During a recent televised encounter with one of their number, I was told that since an eminent astronomer like Sir Frederick Hoyle had said that such an origin of life was impossible, then impossible it was. He reinforced his point with the aid of a Rubik's Cube. If I remember his argument, there was more chance of a blind man solving the scrambled cube than there was of life having come into existence by evolution. What did I think of that, then? he challenged. My first response was that Sir Frederick should stick to astronomy. While this remark was doubtless interpreted as flippant (it was edited from the program), I was making a valid point.

Probability statements have no meaning unless the situation under examination is properly understood. If I take a coin and toss it, I can make a probability statement that there is a 1 in 2, or ½, chance it will come down heads. If the experiment is repeated enough times the results get closer and closer to the predicted probability of ½. I can only make this probability statement because I have a clear picture of the situation: only one side of the coin is a head, the coin is given a good spin each time, and the coin is evenly balanced so that there is no tendency for it to come to rest on one particular side. In other words, the position in which the coin comes to rest is entirely random.

If we know the probability of a single event, we can calculate the probability of that single event occurring several times in succession; it is the product of the individual probabilities. The chances of a coin coming down heads twice in a row is $½ \times ½ = ¼$, that is, 1 in 4. The chances of tossing four heads in succession is $½ \times ½ \times ½ \times ½$, or 1 in 16. When Drs. Hoyle and Wickramasinghe give us a probability statement for the chances of life having originated from non-living materials, they are implying that they can estimate the probabilities of interaction between individual components in the non-living system prior to its becoming a living system. This, in turn, requires a knowledge not only of which particular components came together to form the first living system, but also of their concentrations (the chances of two chemicals interacting increases with their concentration in the solution). The whole argument is patently absurd.

When Dr. Wickramasinghe gave his lecture from the witness stand, he said that his probability statement was calculated on the chances of assembling genes for producing some 2000 enzymes which are found in living organisms (enzymes are proteins that speed up chemical reactions in the body). No one ever asked why he should believe that life could only begin once all of these enzymes could be produced. To ask such a question of a non-biologist would have been pointless, which brings me back to my original remark that specialists should stick to their own specialty.

We will attempt later to show that life probably evolved, step by step, from simple chemical systems all the way to complex molecules, and that this process was directed by selection. Darwin coined the term natural selection for the selection process that occurs in living organisms, and, since this operates through differential breeding success, the term obviously cannot be applied to non-living systems. This does not mean, though, that selection processes cannot occur in non-living systems. We shall see later how such a mechanism can operate between chemical systems that are competing for the same chemicals in their environment.

Dr. Morris gives us a lesson in probability similar to Dr. Wickramasinghe's. He demonstrates the extremely high odds against a simple organism, comprising only a relatively few integrated parts, being synthesized by chance alone. The argument goes like this: Imagine any two components, A and B. In how many ways can they be combined? Obviously only two: AB, BA. Now consider three components, A, B, and C. They can be combined together in six different ways: ABC, CAB, BAC, BCA, CBA, and ACB. In mathematical notation, the number of different combinations of n items is written n! — factorial n. For our last example with three letters, n = 3, and $3! = 3 \times 2 \times 1 = 6$. Citing work done by NASA, Dr. Morris tells us that the simplest type of protein molecule that could be said to be living comprises at least 400 separate amino acids. How many ways can 400 amino acid molecules be lined up, end to end, and joined together to form a protein molecule? The answer is, of course, 400!, which is an exceedingly large number (approximately 6×10^{868}, that is, 6 followed by 868 noughts). Since only one of these combinations is considered to be the "correct" sequence of amino acids, the probability of this being formed by chance alone is 1 in 400!. As the odds are so infinitely high, the event obviously did not occur and life did not arise except by the hand of the Creator.

On first inspection this might appear to be a formidable argument for evolutionists to counter, but, like the other arguments in the creationists' arsenal, this one possesses fatal flaws. In the first place, who ever suggested that complex molecules arose, fully formed, from a mixture of simple molecules? Certainly not evolutionists. A complex molecule, such as a 400-amino-acid protein molecule, represents the last step in a sequence of events that started with the coupling of only a few amino acids. As an evolutionist, I do not visualize that such a complex molecule suddenly appeared on the primeval earth through chance interactions between four hundred amino acid molecules. Instead, I imagine that it was formed by the reaction between an existing and simpler protein molecule and one or more amino acids. This protein, in turn, was formed by interaction between a progressively simpler molecule and amino acids. To use their own analogy, the creationists' strategy is to point to a Boeing 747 and imply that it represents the first flying machine. We would all be hard-pressed to explain how something as complex as a jumbo jet could have been conceived and built from nuts and bolts and pieces of metal without any antecedents.

Another flaw in the creationists' argument is that they overlook the fact that chemicals do not react together willy-nilly, they react together in a selective manner. Some molecules react together readily, others less readily, while still others do not react together at all. The picture that creationists paint of hundreds of simple molecules coming together in one fell swoop, and reacting purely by chance to produce a complex protein molecule, is clearly inconsistent with our knowledge of chemistry.

When creationists challenge evolutionists to explain how life originated, it is somewhat like asking an historian to write a concise account of the history of civilization without having access to any historical documents. We can never know how life originated, but what we can do is to see whether we can reconstruct some of the steps that might have led from non-living to living systems. To do this we need to look at some real evidence. Obviously we cannot duplicate the events that occurred way back in time, but some interesting laboratory experiments have been performed which simulate some of these events. Before evaluating this evidence, though, we have to have some idea of what we mean by a living system.

When we think of things like cats and dogs and trees and tulips, we have no difficulty in recognizing the features they have in common. They all grow, they all absorb nutrients from their environment, they move (though plants are far more limited in their movements than animals), and they all reproduce to leave offspring. As we will see in Chapter 6, we never have any difficulties in distinguishing between things that are obviously different. The problem comes when we approach the transitional zone, and this is true here when we come close to the boundary between life and non-life. What, for example, do we do about things like viruses? These exceedingly small particles, which can only be seen under the high magnification of an electron microscope, comprise a strand of the genetic material DNA (or RNA — ribose nucleic acid — another heredity material) enclosed in a protein capsule. They cannot reproduce themselves, they cannot manufacture enzymes, or any other proteins, not even the protein that encapsulates them. They can be crystallized from solution like an inert chemical substance, they can be desiccated, frozen, and stored for long periods of time, but when they come into contact with the right kind of living cells they start behaving like a living system. They do this by invading a living cell and taking over its internal chemistry so that the cell now manufactures virus particles instead of going about its own business. Viruses are therefore entirely dependent upon living cells. Should we describe them as living or non-living? Most people would probably say that viruses stood on the boundary between the living and the non-living, and their definition of a living system would then include the ability to undergo independent reproduction. This digression into viruses illustrates the point that the transition from non-living to living may not be such a big jump at all.

Before we consider the possible steps leading up to the origin of life on earth, let us say something about the cosmic connection, à la Hoyle and Wickramasinghe. An idea that was popular during the last century, and that has recently undergone something of a revival, is that the earth was seeded with life from outer space. Called panspermia, this idea has two serious disadvantages: first, it merely removes the problem of the origin of life from the earth to some other place; secondly, I have always been puzzled as to how living material was supposed to have survived the transition from space to the earth's atmosphere. If the spores arrived on meteors or some other extraterrestrial bodies, how did they withstand the frictional heat that causes the surface of meteors and spacecraft to glow white-hot? If, instead, the particles rained gently down like dust, how did they survive the ultraviolet and other high-energy radiations above the ozone layer? I believe that panspermia should be left where it belongs, back in the nineteenth century, along with Noah's flood and Adam's rib.

The first thing we have to consider in trying to trace how life began is that the earth is not the same place today as it was four billion years ago. Not only were there no living organisms, but the atmosphere was not the same as it is now. The most abundant element in the universe is hydrogen, which comprises about 92 percent of the total, together with helium, which comprises about 7 percent. The other elements either are present only as traces, or are confined to certain cosmic bodies. It is thought that heavenly bodies, such as our own solar system, are formed by the condensation and subsequent heating of clouds of hydrogen and dust particles. During the heating process, fusion takes place between the hydrogen atoms, and this leads to the formation of the other, heavier elements. From this it follows that the earliest earth atmosphere was composed mainly of hydrogen, probably together with simple compounds such as ammonia, methane, and water. Oxygen was probably absent, and it has been suggested that most of the oxygen we have today has been formed as a byproduct of photosynthesis; when plants convert water and carbon dioxide into sugar during photosynthesis the water molecule is dissociated, releasing free oxygen.

The British biochemist J. B. S. Haldane, writing in the late 1920s, was one of the first to propose that life originated from non-living materials in an oxygen-free atmosphere. This proposal removed the conceptual handicap of visualizing how complex molecules could have been built up in the presence of oxygen. Oxygen, in spite of our biased views of its life-giving properties, is actually a corrosive gas which rapidly breaks down (oxidizes) compounds. The Russian biochemist A. I. Oparin proposed similar ideas and suggested that simple compounds present in the primeval sea (later the term "primeval soup" was coined) might be built up into more complex compounds under the influence of energy sources such as lightning strikes. In the early 1950s, H. C. Urey and S. L. Miller conducted some experiments to see whether this could be achieved in the laboratory in an environment simulating the earliest atmosphere of the earth. Essentially a sterile flask was partly filled with water and connected up to a spark chamber, that is, a large flask fitted with a pair of electrodes across which a continuous spark was maintained. The spark chamber was filled with a mixture of hydrogen, methane, and ammonia, and, by boiling the water in the other flask, the steam generated drove the gas mixture around the apparatus. After several days the solution in the bottom of the flask was analyzed and found to contain more than twenty different compounds, including four amino acids (glycine, alanine, glutamic acid, and aspartic acid). Amino acids, of which there are twenty-one common ones, are the building blocks of proteins. Variations and improvements in the gas mixture, and in apparatus, have since been made by Urey and Miller, and by others, resulting in the synthesis of eight more naturally occurring amino acids, together with adenine, one of the four bases found in the heredity material DNA. Many of the substances formed during these experiments are not found in living systems, but the fact that twelve of the twenty-one amino acids have been formed, and one of the four bases of DNA, is rather impressive.

But are creationists impressed? Not in the least, but their reaction might have been anticipated. In the first place, Dr. Morris does not seem to appreciate that the earth's atmosphere is any different today from what it was in the remote past. He therefore makes the statement that if the evolution of life had in fact occurred, we should still expect to see life being formed from non-living material somewhere on the earth today. This, of course, is just setting up a straw man, because evolutionists recognize that life could not be formed today, simply because it would be immediately destroyed, not only by oxidation but by engulfment by other organisms.

Before leaving the topic, we should mention that organic molecules have been synthesized from simpler ones using methods other than that of Urey and Miller. If a mixture of carbon monoxide and hydrogen is heated in the presence of certain metallic catalysts, such as nickel and iron, various larger molecules are formed. This particular type of chemical reaction, which has important industrial applications, is called the Fischer-Tropsch reaction. Edward Anders and his co-workers, using a modification of this reaction in which ammonia was added to the gas mixture, synthesized nine amino acids and, perhaps more significantly, four of the bases found in DNA and RNA: adenine, guanine, thymine, and uracil. Besides these biologically important compounds, a number of others were formed which are not found in living organisms.

Of particular interest to these scientists was the fact that they identified a similar range of compounds in certain meteorites. (I hasten to point out that I am not trying to make a cosmic connection à la Wickramasinghe and Hoyle.) Anders and his colleagues, who compared the conditions inside their apparatus with conditions pertaining in solar nebulae, concluded that the meteoritic compounds had not been formed by living systems. Their reason was that the mixture contained molecules that are not associated with living organisms — just like the mixture found in the Fischer-Tropsch reaction. If the products had been formed by living systems, they would not be expected to contain these other molecules.

What is the significance of all this? The discovery, in meteorites, of amino acids and bases which are believed to have formed by non-biological processes reinforces the point made by laboratory syntheses that these molecules can be formed by simple chemistry.

However impressed I may be with the results of Urey and Miller's work, and that of Anders and his colleagues, the fact remains that there is an enormous gulf between a mixture of amino acids and some of the components of DNA, and a living organism. How can the gap be bridged? The first problem to resolve is how individual amino acids can be linked together to form long chains. The chemical process of linking small units together to form long chains is called polymerization, and the product is called a polymer. Many of the man-made materials around us today are polymers, including all the various types of plastics. Polythene, for example, is so named because it is a polymer of ethylene. Polymerization takes place inside living cells under the influence of biological catalysts called enzymes. The energy which drives the reaction is derived from the internal chemistry (metabolism) of the cell. But enzymes are polymers themselves; they are proteins, large molecules made up of linked amino acids.

We therefore have an apparent dilemma: how could proteins be formed in the first place if their formation requires the presence of proteins? This is just the sort of chicken-and-egg argument with which creationists love to confront evolutionists. This argument, however, is groundless, because amino acids can be polymerized in the absence of enzymes, and this has been amply demonstrated in the laboratory. There are several simple components, including cyanogen (one of the products formed during the electric-discharge experiment), which facilitate the polymerization of amino acids in the absence of enzymes. The only problem is how to get rid of the water which forms during the polymerization, because this interferes with the reaction. S. W. Fox showed that dry mixtures of amino acids will polymerize within a few hours if heated to about 130°C, and he called the products "protenoids." Polymerization can also be achieved by freezing solutions, so that the water crystallizes out as ice. A third method is to allow the solution of amino acids to be taken up onto the surface of certain clays. Fox's method, which could have occurred when solutions evaporated to dryness in tidal pools, can build polymers comprising in excess of two hundred amino acids.

The building up of proteins is therefore not an insurmountable problem, but we still have to bridge the gap between proteins and self-replicating systems. Obviously some sort of packaging is required to keep all the useful chemicals together and prevent them from drifting off into the "soup." Oparin conducted some interesting experiments with very small droplets of various polymers and carbohydrates, and found that these kept themselves separate from the surrounding water by forming a kind of membrane. These droplets, called coacervates, sometimes behave like living organisms, in that they selectively take up certain substances from the surrounding solution, thereby increasing in size. At a particular point they divide into several daughter droplets and the process of growth and division may be repeated. If certain enzymes are added to the solution, these are taken up by the droplets, which then begin manufacturing products like a small factory.

Nobody is suggesting that these experiments are showing us the actual steps leading up to the origin of life. But they do show us that lifelike behavior, such as the selective uptake of materials from the environment and the ability to reproduce, can be achieved through basic physical chemistry. We should also note that since these chemical systems selectively take up materials from their environment, they are competing with one another. If some droplets can do the job better than others, they will be at an advantage, and when they divide, this advantage is conferred upon their daughter droplets. We therefore have a system comparable to natural selection operating at a level below that of the living organism.

The fact that complex processes can be achieved through basic chemistry is also demonstrated by what can be done with DNA in the laboratory. With the highly sophisticated techniques which are now available, it is possible to slice up pieces of DNA in the test tube and to monitor its self-repair from subunits supplied in solution. The line between non-living and living is a thin line indeed.

Aside from the laboratory experiments which we have discussed, do we have any other evidence to support our ideas of how life may have originated? Obviously there is little point in searching the rocks for a fossil record of early chemical evolution, but we do have some interesting circumstantial evidence in the metabolic processes of living organisms. According to our hypothesis, there was no free oxygen in the early atmosphere of the earth and this means that the earliest life forms were independent of oxygen. We ourselves need oxygen, as do most other animals, and also plants. This is because our cells need it to respire, that is, they need it in order to break food materials down to release energy.

Although most of our cells would die in the absence of oxygen, our muscle cells are able to survive quite happily without it for short periods. This in fact happens every time we do any particularly strenuous exercise, like sprinting for the bus. Because our lungs and vascular system are unable to cope with severe demands for oxygen, the muscle cells have to continue doing their work in the absence of oxygen. They do this by respiring anaerobically (without oxygen), but the amount of energy released by the food (glucose) is much less than during aerobic respiration (breaking down food in the presence of oxygen). This is because the breakdown of glucose is incomplete when oxygen is absent, and we overcome the problem by completing the process after the strenuous activity has been completed. We require oxygen to do this, of course, and during the process we are said to be repaying our oxygen debt. This is why most of us spend so much time puffing and panting after the strenuous exercise has been completed — we are paying off our oxygen debt.

As we will see in Chapter 6, the simplest organisms are those comprising only a single cell. We will also see that not all of these organisms have a nucleus. The nucleus is essentially a small bag in the middle of the cell, enclosed by a membrane, and containing chromosomes. The chromosomes, you will recall, are the threadlike structures which carry the heredity material, DNA. Organisms whose cells have a nucleus are called eukaryotes (*eu* meaning true; *karyen* meaning kernel), while those without are called prokaryotes (*pro* meaning before). We are eukaryotes, as are most other organisms, while bacteria and blue-green algae are prokaryotes. Prokaryote cells are considered to be the most primitive; they are much smaller than eukaryote cells, their chromosomes do not undergo the elaborate process of division, called mitosis, which typifies eukaryote cells, they do not reproduce sexually, and therefore they do not possess the range of genetic variability seen in sexually reproducing organisms. Significantly, many of them can only respire anaerobically, being actually killed by exposure to oxygen. Of those prokaryotes which are able to tolerate oxygen, some are able to survive in its absence, others can only tolerate low concentrations, well below that of the present atmosphere, while others are fully aerobic and die without oxygen. Eukaryote cells all respire aerobically (though some, like muscle cells, can also respire anaerobically), and mitosis cannot occur in the absence of oxygen. The fact that prokaryote cells, which are the simplest, display a range of oxygen preferences, from total intolerance, through partial tolerance, to dependence, is compatible with the hypothesis that the earliest life forms evolved in an oxygen-free world. Furthermore, the similarity between the anaerobic respiration of prokaryote cells and the anaerobic faculty of some eukaryote cells is very suggestive that the former gave rise to the latter. We postulate that anaerobic respiration, or fermentation, was the earliest metabolic process for generating energy and, when oxygen became available through the photosynthetic activities of the earliest plants, that aerobic respiration evolved. Organisms which respire aerobically are at an advantage over the others because this process releases relatively more energy from a given amount of food than anaerobic respiration.

Another piece of evidence to suggest that prokaryotes evolved before eukaryotes pertains to the chemical process called nitrogen fixation. By this we mean the process whereby atmospheric nitrogen is converted into ammonia, and thence into various nitrates. Nitrogen is an essential element of living organisms, but it is only of use when it has been fixed, and only prokaryotes are able to do this. Nitrogen-fixing bacteria, for example, occur inside nodules on the roots of various plants, such as peas and beans, and these used to be planted in the old days of crop rotation (before fertilizers were invented) to enrich the soil with nitrates. Since nitrogen is an essential element for life, it is difficult to visualize how eukaryotes could have existed before prokaryotes appeared.

To summarize the main points of this chapter:

- The numbers games played by creationists to show the improbability of life having evolved are based upon untenable assumptions.
- The earth's atmosphere four billion years ago was rich in hydrogen and devoid of oxygen. By simulating these conditions in the laboratory, it is possible to manufacture a wide range of organic components, including twelve of the twenty-one common amino acids, and four of the bases found in DNA and RNA.
- Contrary to the claims of creationists, amino acids can be built up into long chains ("protenoids") in the absence of enzymes.
- We have a number of pieces of circumstantial evidence to support the view that the prokaryotes were the earliest living cells, and that these gave rise to the eukaryotes. In Chapter 9 we will see that this last point is supported by fossil evidence.

We do not claim to be able to document the evolution of living organisms from non-living ones, but the evidence we have presented makes a convincing case that such an evolution was possible.

FIVE   *Noah's Ark: Fact or Fable?*

 I FIND IT HARD TO BELIEVE that anyone could accept a literal interpretation of the Noachian flood. I find it equally hard to believe that I should be sitting at my desk in 1982 documenting the reasons why the flood could not have happened according to Genesis. The fact remains, though, that a large number of people do believe that a worldwide flood destroyed all living things on the earth, save those that were taken aboard the ark. These same people believe that fossils are the remains of those organisms that perished in the flood, and that their relative position in the rocks is largely a result of where they were living when they were overcome by the flood. If people want to accept the story of the ark on faith, I have no wish to take issue with them. However, if they tell me that they have scientific evidence to support their belief, as Dr. Morris does, it is an entirely different matter. Dr. Morris tells us that

> If the system of flood geology can be established on a sound scientific basis, and be effectively promoted and publicized, then the entire evolutionary cosmology, at least in its present neo-Darwinian form, will collapse. This, in turn, would mean that every anti-Christian system and movement (communism, racism, humanism, libertinism, behaviorism, and all the rest) would be deprived of their pseudo-intellectual foundation.

After firing this broadside against anti-Christians, a term which he broadly interprets as all those who are not creationists, Dr. Morris gives us a few facts about the flood:

- The flood was worldwide and, since the ark came to rest on Mount Ararat (Genesis 8:4), the waters were at least 5000 m deep (about 3 miles) based on the present elevation of Mount Ararat.
- The flood lasted for a little more than ten months (after the forty days and nights of rain).
- The ark had a capacity at least equivalent to 522 standard railway stock cars. This, according to Dr. Morris, is more than twice the volume required to accommodate two of every species of land animal that ever lived.
- The flood destroyed everything; it "overturned the earth."

Before we examine the plausibility of these points, we should add two more pieces of information from Genesis that Dr. Morris has omitted.

- Noah was instructed to take seven of every clean beast (animals with a cloven hoof that chew the cud, that is, ruminants) and seven of every bird, as well as two of every other kind of animal.   (Genesis 7:2–3)
- Noah was instructed to take enough food for all the animals, as well as for his own family.   (Genesis 6:21)

We should also add that although Noah was not instructed to take marine organisms aboard the ark, the fact remains that most of these, from sea anemones and corals to sharks and whales, are unable to tolerate fresh water. If the forty days and nights of rain covered the earth with 5000 m of water, it follows that most marine organisms would have perished, unless they were taken into the ark. (The creationists might argue that the waters from the "fountains of the great deep" maintained the salinity of the sea, but this explanation strays beyond the bounds of science.)

The first and obvious question is whether the ark could have carried all of its occupants. According to Genesis 6:15, the ark was 300 cubits long by 50 cubits wide and 30 cubits high. A cubit is about 44 cm; therefore the dimensions of the ark were approximately 150 m × 25 m × 15 m, which is an overestimate (I am intentionally erring on the side of the creationists). The maximum capacity was therefore 56 250 cubic meters, which is again an overestimate because it does not allow for the fact that the ark would have been boat-shaped and not rectilinear, or for the space taken up by the walls that Noah was instructed to build (Genesis 6:14), or for any of the superstructure and decking, or for the barriers and cages that would have been required to keep the various animals apart. The available space would have been considerably less than 56 250 cubic meters.

There are over one million species of animals living today, together with about half a million plant species (the latter could, of course, for the most part have been taken along as seeds). Although the exercise is rather simplistic, let us now divide the 56 250 cubic meters of space, which we know to be a gross overestimate, by the 1.12 million number of animal species to see how much room each species could occupy. The result is 0.05 cubic meters, which is about one-third of the capacity of a domestic oven. Now this is more than enough space for a pair of ants or a pair of mice but it is rather inadequate for a pair of squirrels, and quite out of the question for a pair of zebra. However, we have to bear in mind that most animals are smaller than zebras; insects, after all, account for about seventy-five percent of the total number of species, and most of these could live in a space of 0.05 cubic meters quite happily. If we therefore allocated the space according to the size of the animals, we might conceivably pack them all into the space, but we have to remember that this space is not available space, merely the volume of the ark. It is therefore most unlikely that the ark could have held all the animals.

What about the food for the journey? As we saw in Chapter 3, warm-blooded animals consume large quantities of food. An African elephant, for example, eats about 160 kg of fodder a day, so that a pair of them would consume about 96 metric tonnes during the voyage. Now 96 tonnes of hay occupies a considerable amount of space when you consider that a standard bale weighs about 32 kg. Just imagine how much hay would be needed for all the other pairs of ungulates — hippo, tapir, zebra, warthog, llama, boar, peccary, ass, camel, chevrotain, horse, to mention only a few. But we have forgotten something: the ruminants were not in pairs but in sevens — seven pronghorns, seven bison, seven muskox, seven ibex, seven moose, seven sheep, seven wapiti, seven goat, seven gazelle, seven eland, seven wildebeest, seven oryx, seven impala, seven reindeer . . . They would eat an awful lot of hay in three hundred days, and think of all the fertilizer Noah would have to shovel away. Then there are all those carnivores to worry about. A lion, for example, consumes its own weight in food every eight or nine days, so that a pair of lions would eat about

35 kg of meat a day. This would amount to some 10 tonnes during the voyage, and what about all the other pairs of carnivores? Aside from the space problem of storing many hundreds of tonnes of meat, there is the problem of keeping it fresh. Remember that this was before the days of refrigerators. I suppose the meat could have been salted, but I am not sure whether carnivores will eat salt meat, and if they did, it would make them very thirsty. Now this raises another problem.

We have touched upon the problem of food storage but have said nothing about fresh water. Would the flood waters have been drinkable with all those dead things floating around in it, or would Noah have had to provision the ark with water? If fresh drinking-water had to be carried, this would have presented horrendous storage problems.

What about the birds? There are over 8000 species, and Noah was told to take along seven of each kind. Fifty-six thousand birds would eat their way through a mountain of food during the voyage. There is also the problem of animals with special diets. Koala bears, for example, will only eat eucalyptus leaves, and these have to be supplied fresh every day. I can see no alternatives but having eucalyptus trees growing on board.

And what about the special requirements for the marine organisms? Just consider how many watertight containers would be required to keep all the various sea creatures in. Imagine how large a tank would be required for just one species of whale, let alone for all the others. The sharks would need a big tank, too, and so would the seals, and the sea lions and the walruses, and the squids and the octopuses, and all the other large animals that live in the sea. And they would have to have been kept separate from each other; otherwise they would have started eating one another. There is also the problem of keeping the sea water clean and healthy; not an easy task, as anyone who has tried to keep a marine aquarium knows.

The only way for the creationists to remove the logistics problem of the ark is to argue that if God willed it to be so, then so it was. But this strategy defeats Dr. Morris's wish for the Noachian flood to be "established on a sound scientific basis."

Not only are the packaging and maintenance problems of the ark untenable, but so too is the problem of dispersing the animals and plants to their native lands at the end of the voyage. We are told in Genesis 8:4 that the ark came to rest on the mountains of Ararat. How did all the animals find enough to eat on that mountainside in Turkey when the flood subsided?

We are told that everything was destroyed during the flood; it would therefore have taken some time before plants could have started growing again, and these presumably would have to have been sown by Noah. What did the plant-eaters do while they were waiting for the plants to grow? What did the meat-eaters do while they were waiting for the plant-eaters to multiply? Could Noah have had sufficient supplies in that first season? Hardly — we have already shown that he could not have carried anywhere near enough supplies for the original voyage.

Assuming that all the animals managed to survive that difficult period following the flood, which is seemingly impossible, how did they disperse themselves to the four corners of the earth? How did the two pandas make it right across Eurasia to China? How did the two kangaroos hop across Eurasia and then across the ocean to Australia? Are we to suppose that it was not the original Noachian pairs that traveled these enormous distances, but later generations? If that is so, then we have to accept that Turkey and its environs must have been suitable for all the animals and plants; otherwise they would all have perished right there and then. Can you imagine a climate that would have been suitable for tropical and for polar organisms at one and the same time? The story of the ark, quite clearly, cannot stand up to scientific scrutiny. This is hardly surprising, as I do not believe it was ever meant to be taken literally.

The other aspect of the Great Flood that raises grave problems for the creationists is the order in which the animals and plants came to be deposited in the sediments that settled out from the tempest. Dr. Morris tells us that there are a number of obvious predictions that could be made regarding the stratigraphic levels and abundances of various groups in the fossil record. These predictions are as follows:

1. Marine invertebrates would tend to predominate, since there are many more of them, and, being relatively immobile, they would usually be unable to escape.

2. Animals living in the same ecological niche would tend to be buried together.

3. Animals living at the lowest elevations would tend to be buried at the lowest levels, so that the relative position of fossils in the geological column would correspond to the elevations at which they lived.

4. Marine invertebrates would tend to be found at the lowest

levels in any local geological column because they live on the sea bottom.

5. Marine fishes would be found at higher levels than bottom-living invertebrates because they live at higher elevations, and could escape burial longer.

6. Amphibians and reptiles would tend to be found at higher levels in the column than fishes, enclosed in sediments at the interface between land and water.

7. There would tend to be few, if any, terrestrial sediments, or organisms, in the lower levels of the column.

8. The first appearance of land plants in the column would tend to be on the same level as amphibians and reptiles, when rafts of lowland vegetation were carried down to the seashore by swollen rivers.

9. In marine strata the invertebrates would tend to settle out according to size and shape and would therefore form assemblages of similar-sized and shaped fossils. The simpler animals, being more nearly spherical or streamlined, would tend to settle first, because of their lower drag in the water.

10. Mammals and birds would tend to be found higher up in the column than reptiles and amphibians because of both their habitat and their greater mobility. Few birds would be found.

11. Because of their herding instinct, higher animals would tend to be found together in large numbers, if at all.

12. Higher animals (land vertebrates) would also tend to be distributed in the geological column in order of their size and complexity, because of the greater ability of the larger and more diversified animals to escape burial for longer periods.

13. Few human fossils would be found. People would tend to escape burial, and after the water receded, their bodies would lie on the ground and decompose.

Dr. Morris adds a cautionary note that although these predictions are expected statistically, some exceptions would be expected due to the cataclysmic nature of the flood.

This is a formidable list, but we will work our way through the points, taking them in the order in which they are given. In each case we will first comment on the logic of the prediction, then give our own prediction based upon conventional steady-state geology. Lastly, we will see how the predictions are supported by the fossil record.

1. The sea covers about seventy percent of the surface of the earth, and invertebrates far outnumber vertebrates, so the prediction that marine invertebrates should predominate in the fossil record seems reasonable. We must point out, though, that most marine organisms live in the shallow waters around the coasts, and this comprises only a fraction of the total area of the sea. Furthermore, the insects are the most abundant of all animals, certainly in terms of the total number of species (with about 850,000 species, they comprise about seventy-five percent of the total number of living animals) and probably also in their biomass (total mass or all the individuals). In terms of numbers, then, the creationists should probably have chosen the insects as being the most numerous fossils.

   The second part of the prediction, that the marine invertebrates, being relatively immobile, would be unable to escape, seems illogical. From what would they be trying to escape? Obviously not from drowning, because they are already living in the water; however, most of them would have liked to escape from the dilution effect of the rainwater. This is because most marine organisms are intolerant of lowered salinity — they tend to swell up, burst, and die. However mobile an aquatic organism might have been, it would have been unable to escape from this dilution effect.

   I would predict that marine organisms, especially shelled forms like molluscs, would be particularly plentiful in the fossil record for three reasons:

   • Sedimentation rates in coastal waters (where most invertebrates are found) tend to be higher than elsewhere, and high sedimentation rates, as we will see in Chapter 8, increase the chances of preservation.

   • Shells, being robust, stand a better chance of preservation than many other structures.

   • Molluscs are abundant; in terms of numbers of species they are second only to the insects.

   When we look at the fossil record we find that marine invertebrates do tend to predominate and that molluscs are among the commonest of all fossils. Creationists and evolutionists therefore make the same prediction, which is supported by the fossil record; but creationists are not completely sound in their reasoning (they should have chosen insects).

2. Animals living in the same ecological niche, for example animals living under rocks (woodlice, beetles, slugs, snails), or animals living in trees (squirrels, birds, caterpillars, weevils), would not be expected to be buried together in the aftermath of a flood. When floods sweep across the land, churning up the soil, rolling over stones, and knocking down trees, animals and plants are mixed up and tumbled along irrespective of where they were living. The creationists' prediction that organisms would be preserved in communities is therefore totally illogical. Only under the steady-state conditions predicted by evolutionists would animal communities tend to be preserved, and this is what we usually find in the fossil record. Creationists are therefore correct in their prediction but for completely wrong reasons.

3. We are told in Genesis 7:12 that it rained for forty days and nights, which is an extremely short period of time for the floodwaters to rise up and cover the highest mountains. From this it would seem that the elevation at which organisms were living at the time of the flood would have been of less significance to their final resting place than how mobile they were and how well they floated. The creationists' prediction that organisms would be found in the geological column in the order of the elevation at which they lived is therefore illogical. In any event, the stratigraphic level of organisms in the geological column bears little relationship to the elevation at which they lived, how mobile they were, or how well they may have floated. We will document this when the relevant groups are treated in the creationists' list of predictions.

4. The claim that marine invertebrates would tend to be found at the bottom of the geological column because they live at the bottom of the sea is founded upon a basic misunderstanding of how marine organisms are distributed. Relatively few marine animals, invertebrate or vertebrate, live in the depths of the sea, simply because there is very little to eat in those cold and dark places. Very little light penetrates beyond about 100 meters, so there is no photosynthesis beyond that depth and therefore there are no plants; the only nutrients reaching the depths are the leftovers and dead remains of animals living in the upper layers. The vast majority of marine organisms live in shallow coastal waters, and these comprise the smallest fraction of the total area of the sea, no more than about five percent. If crea-

tionists knew anything about the distribution of marine organisms, they would predict that marine fossils would be abundant at the level of the interface between land and water. They would certainly not predict abundant fossils at the lowest levels of the column, which they interpret as being on the sea bottom.

Notwithstanding the fact that the creationists' prediction is founded upon a misconception of animal distribution, do we find marine invertebrates at the bottom of local geological columns, or can they also be found at the top?

Anyone who has spent any time looking in marine sediments for fossils knows that marine invertebrates can be found at any level in a local geological column. For example, if we took a stroll along the beach at Lyme Regis, in southern England, one of the most celebrated localities for marine fossils, we should find invertebrate fossils at all levels, from the beach all the way up to the top of the cliffs. If we were lucky enough to find remains of some of the vertebrate fossils which have made Lyme Regis famous throughout the world, these would be found at any level too. There is, therefore, no tendency for the invertebrate fossils to be found at the bottom of the column.

5. The creationists' prediction that marine fishes would be found at higher levels than bottom-living invertebrates, both because of their higher elevation and because of their ability to escape burial longer, seems reasonable. As an evolutionist, I too would expect fishes to appear higher in the geological column than invertebrates, but for different reasons (discussed in Chapter 8). This prediction, though, only applies to the first appearance of fishes; once fishes had appeared in the record, there is no reason why their order with invertebrates should not be reversed. Thus, while the earliest fish skeletons (Ordovician in age) would not be expected to appear in the geological column below the level of invertebrates, there is no reason why, say, Devonian fishes should not be found below the level of invertebrate fossils. According to the creationists, though, fossil fishes would always be found above the level of bottom-living invertebrates.

How do these predictions match up with the fossil record? As we will see in Chapter 8, the earliest fossils are invertebrates, which satisfies both the creationists' and the evolutionists' prediction. After the first appearance of fishes, though, we find that the order of fishes and invertebrates is frequently reversed; this is contrary to the creationists' predictions but corresponds to that of evolutionists. Large numbers of fossil fishes, for example, have been collected from rocks of Devonian age, and these are much lower down (older) in the geological column than the Chalk, which is noted for its invertebrate fossils.

6. The prediction that amphibians and reptiles would be found at higher elevations than fishes — at the interface between land and water — is based upon the simplistic and erroneous belief that amphibians and reptiles live only at sea level. Dr. Morris presumably has never ventured into the mountains of his native California; if he had, and if he had been observant, he would have seen all manner of amphibians and reptiles there — salamanders, toads, frogs, tortoises, lizards, and snakes. Tree frogs, which are found in mountainous rain forests in many parts of the world, are often found at high altitudes, and most of the giant tortoises I saw when I visited the Galapagos were at the highest altitudes, living in the lush vegetation that grows around the rim of the volcanoes there.

As in the previous case, creationists and evolutionists have similar predictions for the fossil record, but these are for completely different reasons. Once again the evolutionists' predictions only apply to the first appearances of the groups. I would expect the first amphibians and reptiles to appear at a higher level than the first fishes, but subsequent amphibian and reptilian fossils should be found both below and above the levels that yield fish fossils. Creationists, however, would predict that amphibians and reptiles would always be found at higher levels than fish fossils. When we look at the fossil record we find their predictions are not supported. For example, one of the richest localities for fossil fishes is the Green River shales of Wyoming, which are Eocene in age, and these lie at a higher level than the famous dinosaur quarries of the western interior.

7. The prediction that there would be few, if any, land sediments, land plants, or land animals in the lower levels of the geological column is based, presumably, on the assumption that land, and land organisms, are found only at high and not at low elevations. One has only to think of places like East Anglia in England, or of the Netherlands, where much of the land is below sea level, or of inland bodies of water like Lake Titicaca in Peru, which is almost 4000 m above sea level, to realize that land is not necessarily found

at higher elevations than water. The creationists' prediction is therefore not on a firm foundation. Evolutionists would predict that land and aquatic sediments would be found in any order in the geological column. Do we find land sediments and land plants and animals in the lower levels of the geological column, or only at higher levels as creationists predict?

Land plants are certainly present in rocks of Silurian age, which is low down in the geological column, and the Permian, which is still not very high up, is noted for its terrestrial reptiles (the mammal-like reptiles). This is not in accord with the creationists' prediction. Furthermore we see various alternations of land and aquatic sediments throughout the column, which is in accord with the evolutionists' predictions.

8. Dr. Morris's prediction about the coincidence of the first land plants, amphibians, and reptiles in the geological column, due to rafting, is founded upon the misconception, discussed above, that amphibians and reptiles live close to sea levels. Notwithstanding the illogical foundation of the prediction, is it confirmed by the fossil record? Absolutely not. The first land plants appear in the Silurian, but amphibians did not appear until the succeeding Devonian Period, and reptiles not until the Carboniferous Period which followed.

9. The invertebrate fossils, the creationists say, would tend to be sorted in marine sediments according to their size and shape, the spherical ones settling at the lowest levels because of their lower drag. This seems a reasonable prediction if one believes in the flood, though I wonder why it should only apply to marine invertebrates and not to all organisms. The flood, after all, was said to have swept right across the land, so all organisms should have been subjected to the same hydraulic sorting processes. This is a minor point, though. The important one is whether the prediction is supported by the fossil record, and anyone who has taken the time to do any fossil-hunting knows that it is not. I should like to take Dr. Morris for a stroll along the beach at Lyme Regis and let him see the size diversity in the invertebrate fossils for himself. The commonest fossils there are the ammonites, and these coiled shells range in size from a few millimeters to more than one meter. Sometimes the largest ones are found below the level of the smaller ones, sometimes the reverse, and sometimes they are intermingled at the same level.

10. The prediction that fossils of mammals and birds would tend to be found at a higher geological level than those of reptiles and amphibians is based upon misconceptions of habitat preferences similar to those in predictions 4 and 6. Aside from the fact that many groups of mammals and birds spend all or most of their lives at sea (whales, dolphins, sea lions, seals, sea cows, otters, penguins, cormorants, and pelicans, to name just some), the suggestion that mammals and birds tend to live at higher altitudes than reptiles and amphibians is just not true. Furthermore, the contention is not supported by the fossil record, because, although the first amphibians and the first reptiles occur at lower horizons than the first mammals and the first birds, the order is not maintained for later representatives of these four groups. Mesozoic mammals and birds, for example, are found at a lower level than Tertiary reptiles and amphibians.

11. The prediction that higher animals would tend to be found together in large numbers in the fossil record because of their herding instinct overlooks the fact that there is no correlation between herding instinct and complexity. Is the cheetah, a solitary animal, lowlier than the lion? Are colonial bees, ants, and locusts higher than solitary sharks, eagles, and leopards?

    Does the fossil record support this unfounded prediction? I would describe the ichthyosaurs, a group of marine reptiles that lived during the age of dinosaurs, as "higher" vertebrates, and they are often found together as fossils, but so too are fishes, which would be described as "lower" vertebrates. The primates (see Chapter 14) are among the "highest" animals in terms of their intellect, and most of them live together in communities, but their fossil remains are seldom found in large numbers, solitary finds being the rule rather than the exception.

12. The "higher" vertebrates, the creationists say, would tend to be distributed in the geological column in the order of increased size and complexity, reflecting their ability to escape burial for longer periods. Does this make any sense at all? Would, say, the albatross, which we know from observation stays on the wing throughout the year, only coming onto land to breed, be less able to escape burial than the larger and more intelligent apes? I think not.

    Is there any support for this illogical prediction in the fossil record? As we know, there is a sequential appearance of more complex forms in the geological column, but this applies to the first appearances of groups. As we have said before, there are plenty of fossil birds and mammals that occur at lower levels than fossil reptiles.

13. The prediction that there would be few human fossils because "Men would escape burial for the most part and, after the waters receded, their bodies would lie on the ground until decomposed" does not seem altogether logical to me. If the land were completely covered with water, as stated in the book of Genesis, then all humans would have eventually been swept into the water, even those who had managed to escape the rising water by climbing to the highest points. Drowning victims, as every forensic scientist knows, sink to the bottom soon after death, and it would therefore be expected that the victims of the Great Flood would stand a good chance of burial and subsequent fossilization.

Except for relatively recent material, human fossils are fairly rare, and while this is in accordance with Dr. Morris's predictions, it is contrary to what I would predict if the flood had been a reality. As an evolutionist, I would predict that fossil man would be fairly rare, but this is because he lived in environments that do not readily form fossils.

If there had been a great flood which swept right across the land, uprooting trees and sweeping everything along in its path, would we not expect things to have settled on the bottom according to how well they floated? Would we not expect to find a veritable jam of uprooted trees and the like at the top of the column? We would at least expect to find a predominance of trees in the upper layers — but we do not. We would also expect to find the densest organisms, like corals and shells, down at the bottom of the column, but we do not. We would expect to find the sediments graded according to particle size, just as they are on the beach. The largest boulders should be found at the bottom, grading into pebbles, coarse gravel, sand, and then mud and clays at the top. We do not see this, though, and the various sediments occur at all levels in the geological column.

Something that has always puzzled me about the Great Flood is where all the water went afterwards. It would have taken an awful lot of water to completely cover the highest mountain peaks, and all that would have to have gone somewhere. Could it have become locked up in the polar ice caps? Hardly — they only account for about two percent of the world's water (fresh and salt), and it has been estimated that if all this ice melted, sea levels would not rise by more than about 40 m. Could it have seeped down through cracks in the earth's crust? This does not seem very likely considering how hot it gets down there. Perhaps the water all evaporated and went up into the atmosphere. Yes, that must be it, that explains why it rains so much!

SIX

Gaps or Continuity? A Look at the Living World

ACCORDING TO DR. MORRIS, the gaps between major groups of organisms, both living and fossil, are real, and serve to show that the different kinds of plants and animals have been created separately. "There is no evidence that there have ever been transitional forms between these basic kinds," he states. He reached this conclusion largely by assessing the fossil record, but this tells only part of the story. Perhaps creationists are swayed by the popular misconception that we can trace evolutionary relationships only by studying fossils, but this is illogical. Suppose, for example, that two particular groups of living organisms are connected by (living) intermediate forms, so that there is no gap between them. We would conclude from this that the two groups have always been connected throughout their fossil record. The alternative is that the connection appeared at some later date, which is highly unlikely. Let us also remember that stasis (no evolutionary change) is perfectly acceptable to evolutionists and we have no reason to believe that, say, a modern insect or a modern worm should be significantly different from a Devonian (350 million years ago) insect or worm. Living organisms provide us with a wealth of evidence that major groups are interrelated, as the present chapter will show.

There is much to be said for the classical way biology was taught when I was a student. As a zoologist I had to study all types of animals, both living and extinct. This included a wide array of single-celled blobs of life, all manner of creeping and crawling and slippery things without backbones (invertebrates), and an impressive array of vertebrates. The theme and structure of the course was systematics (the study of classification and relationships), and while most of the beasts we studied fell nicely into one category or another, there were those that refused to co-operate. At that time I viewed such organisms with some disdain; they just messed everything up and I was not interested in the fact that they showed evolutionary relationships between particular groups.

My first surprise as a purist zoology student was that there was no satisfactory way of separating the plants from the animals. The theoretical distinction between the two is the way of feeding; plants manufacture their own food from the air and water using light energy (the process called photosynthesis), while animals obtain their food by eating plants or other animals. Among the countless single-celled organisms that can be seen in a droplet of water — *Amoeba, Ceratium, Paramecium,* and *Euglena* to mention but a few — there are some that manufacture their own food and others that consume ready-made food. There is no convenient way of drawing the line. There are even some types, such as *Euglena,* of which some of the species are plants, others are animals, and others still can be plant or animal depending upon their environment. This is a most perplexing situation for people interested in classification, and some of them have spent considerable amounts of time trying to decide how to deal with the problem. The modern solution is to classify all single-celled organisms (organisms that have a well-defined nucleus) into one kingdom, the Protista, recognizing that there is a continuous spectrum between plants and animals. This, of course, flies in the face of Dr. Morris's claim that there are no transitions between basic kinds — what could be more basic than plants and animals?

As if having no clear-cut distinction between plants and animals were not bad enough, this same group contains organisms that transcend the line between single-celled and many-celled (or multicellular) organisms. If you scraped off some of the green film that collects at the top of a cold-water fish tank and looked at it under a microscope, one of the organisms you would probably see is a protistan called *Gonium. Gonium* comprises four round green cells enclosed in a thin capsule. The four cells are probably best thought of as forming a colony rather than a multicellular organism because each one looks like its neighbor and each performs the same function. Each, for example, has a pair of flagella (whip-like processes) that project through the capsule and are used for swimming, and each cell manufactures food through the process of photosynthesis. There is, therefore, no division of labor or specialization of any of the cells for separate functions that characterize the multicellular organisms. But our sample from the fish tank might also contain a specimen of *Volvox*, a rather beautiful sphere filled with hundreds of emerald-green cells embedded in a jelly which they secrete. There may be some connective strands running between the cells and there is a degree of specialization among them because some are modified for reproduction. Should this complex be described as a colony of single-celled organisms or as a multicellular organism? Perhaps it should be viewed as an intermediate stage.

There are many more organisms which straddle the line between the unicellular and the multicellular levels of organization. Some, like *Volvox*, manufacture their own food by photosynthesis and are therefore plants, but there are others that feed on ready-made foods and are therefore described as animals. A particularly problematic group of protistans to classify are the slime molds. For most of the time they look like amoebae — shapeless blobs of protoplasm that move with a slow, flowing motion, engulfing bacteria and other particles of food as they go. Because the individual cells stay joined together when they multiply, the organism is essentially multicellular, but it cannot be described as a truly multicellular organism because there are no boundaries between the cells. Under certain conditions the organism develops one or more stalked balls, very much like the fruiting bodies of fungi, and each ball comprises a number of individual cells which are specialized for dispersal. The organism can therefore be said to be multicellular at this stage of its life cycle. When these individual cells, or spores, are released, they grow flagella which they use for swimming. They eventually fuse together in pairs, the flagella are lost, and a new amoeba-like organism is formed. The cycle starts all over again. In this organism, then, we see a number of transitions: from flagellated cell to amoeboid cell, from unicellular to multicellular organism.

The organisms we have been considering so far, which are united within the kingdom Protista, are characterized by having a distinct nucleus. This, as we saw in Chapter 3, is essentially a thin-walled structure within the cell where the genetic material is kept. Although protistans are relatively simple organisms, they are not the simplest, for there is another large group of unicellular organisms, placed within the Kingdom Monera, which do not have a nucleus. They are of interest to us here not only because they show transitions between plants and animals, but also because some transcend the boundary between the unicellular and the multicellular levels of organization.

There are two basic types of monerans, bacteria and blue-green algae. Blue-green algae can be found in ponds and in wet soil, and some even thrive in hot springs and thermal pools, where they form brightly colored mats upon the rocks. This is truly remarkable, because the water in these pools is not far from boiling-point. Most of the blue-green algae can produce their own food by photosynthesis and are therefore classified as plants, but others absorb food, in solution, from their surroundings, and would therefore be classified as animals. Most bacteria are unable to make their own food and, like some blue-green algae, absorb nutrients in solution from their surroundings, and in this regard they could be classified as animals. Many bacteria are parasitic, living inside the bodies of other organisms and feeding at the host's expense. They are often harmful to the host because of the poisonous materials they manufacture as waste products. Both bacteria and blue-green algae occur as single cells, but many of them are able to form colonies which have the form of chains or clusters. The cells maintain their separate identities and all do similar jobs, but in some blue-green algae there is actually a division of labor among the cells and the whole colony has the makings of a simple multicellular organism.

This brief encounter with simple organisms, which gives us some idea of what the earliest life-forms were like, establishes two points: (1) there is no clear-cut line between plants and animals, and (2) there is no clear-cut distinction between single-celled and many-celled organisms.

What do the creationists have to say about all this? Actually they seem to spend all of their energies in the valiant quest for fossil evidence of these minute blobs of life, but Dr. Morris does make some reference to living forms, saying that "if the evolution model were valid, one would expect to find a horizontal continuum of living organisms, rather than clearcut categories." If I interpret this statement correctly, I believe that we evolutionists have just scored a point.

An experiment that has always intrigued me involves a

sponge and a piece of silk. Sponges, not the imitation plastic variety, but real sponges, are quite common shore animals. They are multicellular, but are fairly simple in that they have only three main types of body cells. One of these types, which has a flagellum surrounded by a thin collar, looks much like a certain type of unicellular organism, and functions to capture food. The other cells are relatively unspecialized and make up the sac-like body. Apparently if you chop and grind up a sponge and filter it through silk to break it up into its individual cells, life goes on! The flagellated cells go on feeding, now as independent operators, but they begin bunching up with other cells to form small colonies. These small colonies eventually organize themselves into new, but small, sponges. This experiment graphically shows what an elastic line there is between unicellular and multicellular organisms.

Worms are not very musical. Darwin, we are told, tried playing a tuba in front of one to see whether he got any response. The worm in question remained unmoved by the performance. Aside from being unmusical, worms are long and thin, and their bodies are soft and divided into numerous segments. They have no antennae and no legs, though some worms, like the ones used as bait for sea fishing, do have tufts of bristles projecting from the sides of the body. Each body segment has a pair of tufts and these assist their swimming and burrowing activities. Because they breathe through their skin (many have external gills), the skin has to be kept moist; consequently they are restricted to living in moist places, like the soil and the sea. They lack a hard skeleton and maintain their body shape by the pressure of their body fluid. Their skin has a protective outer layer, called the cuticle, but this has to remain thin to allow for breathing.

Insects and their relatives represent a big evolutionary leap forward from the worms. Like the worms they have a segmented body, but this is not soft and moist, being instead covered with a thick and stiff cuticle which fits them like a suit of armor. They wear their skeleton on the outside, whereas we, and the other vertebrates, have our skeletons on the inside. We have probably all handled these animals at some time or other — beetles, crabs, and lobsters to name just a few — and know how hard they feel. In order for them to be able to move, their armor has to be jointed, and this is why the group is called arthropods (meaning jointed legs). One of the problems of living in a suit of armor is that it restricts growth, and arthropods have to shed their outer skeleton periodically when

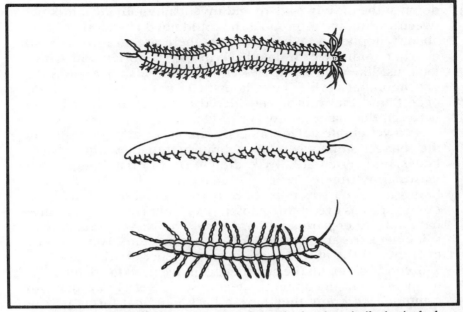

A worm, an onychophoran, and an arthropod, showing similarity in body form.

they are growing. Also, because the cuticle is not porous, they cannot breathe through it. Insects have overcome this problem by a complex system of tubes, called trachea, which convey air from the surface to the living cells of the body. The arthropods are a biological success story, far outnumbering all other animals with almost one million species. Worms, in contrast, number only seven thousand species.

Are arthropods related to worms? Actually we have good evidence that they evolved from them, the gap being bridged by a small group of animals that live their lives unobtrusively beneath the bark of dead trees in the warmer parts of the world. These animals have no common name and are called onychophorans. They are fairly small, about 10 cm long, and have an elongated and internally segmented body reminiscent of a worm. Unlike worms, however, they have many pairs of legs, and a pair of antennae on the head. I imagine that most people would have difficulty deciding whether they were looking at a worm or at some sort of arthropod if they saw an onychophoran in their back garden. Like worms, they have a thin cuticle, and are therefore soft to the touch, but they breathe through a tracheal system just like insects. If an evolutionist had to sit down at the drawing-board and invent a hypothetical link between worms and arthropods, he could not do better than draw an onychophoran. What is more, there are beautiful fossil onychophorans which date back to the Cambrian and which look just like their living descendants. Dr. Gish overlooks the onychophorans when he tells us that "not a single fossil has been found that can be considered to be a transitional form between the major groups, or phyla."

As a vertebrate paleontologist I am especially interested in the connection between invertebrates (animals without backbones) and vertebrates. According to Dr. Morris, though, I am wasting my time: "The evolutionary transition from invertebrates to vertebrates must have involved billions of animals, but no one has ever found a fossil of one of them. Invertebrates have soft inner parts and hard outer shells; vertebrates have soft outer parts and hard inner parts — skeletons. How did one evolve from the other? There is no evidence at all."

Before picking up the gauntlet again we need to spend a few moments thinking about the features that vertebrates have in common. We know they have a backbone that forms part of their internal skeleton, and that this is not always bony, being made of gristle (cartilage) in sharks and their allies, and also in the lampreys. All vertebrates at some stage during their development have a stiff rod, called a notochord, running along the back, but this is retained in the adult only in the most primitive members, for example in the lampreys. All vertebrates have a hollow central nervous system (brain and spinal cord) and this runs along the back (described as dorsal in position), above the gut, or digestive tract, and closely associated with the backbone. This contrasts with invertebrate animals, whose (solid) nerve cord runs along the front (described as ventral in position), below the gut. The more primitive vertebrates, which includes all the fishes, have gill slits, but we can see precursors of these structures during the embryonic development of all vertebrates. I hasten to point out here that I am not advocating the old idea that the evolutionary history of an animal is recapitulated during its embryology (see Chapter 10). I am, instead, pointing out that the ability to form gill slits has been inherited by all vertebrates, but this ability usually only shows itself during early embryonic life.

There are many more characteristics that vertebrates share, but we need only add one more for our present purpose: a segmented body. This does not mean, of course, that our bodies look segmented on the outside like an earthworm's, but

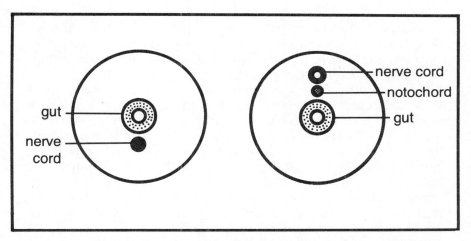

A diagram of transverse sections through an invertebrate (left) and a vertebrate.

we are internally segmented — just think about the bones in our back and the paired muscles associated with them, not to mention our ribs. At this point it might be helpful to list the vertebrate characteristics we have just considered:

- internal skeleton
- notochord at some time during life
- hollow dorsal nerve cord and brain
- gill slits at some time during life
- segmented body

In 1774 the German zoologist Pallas published an account of a small animal that had been collected off the Cornish coast of England. The creature was small, only about 3 cm long, and had an elongated body, flattened from side to side and pointed at each end. Pallas was apparently unimpressed by the new creature, describing it, erroneously, as a slug. Later workers, however, got very excited about it when they discovered that it had a number of features in common with vertebrates. It had a notochord, a hollow dorsal nerve cord that is even expanded at the front end into a rudimentary brain, gill slits, and a segmented body. As it happens, this little animal, usually called amphioxus, turned out to be quite common, and is found in oceans around the world. It spends most of its time burrowed in the sand with only its front end projecting, and feeds by straining food particles from the water using an elaborate sieve mechanism. Though fairly inactive, it does swim freely from time to time.

Because it lacks a vertebral column, amphioxus cannot be described as a vertebrate, but it is obviously closely related to vertebrates, having more things in common with them than with any of the invertebrate animals. It is accordingly classified within the same major group as the vertebrates, the Phylum Chordata (meaning having a notochord), along with a number of other animals that have an affinity with the vertebrates.

How should we interpret amphioxus? I regard it as a surviving member of a group of organisms from which the vertebrates evolved. I am not suggesting that amphioxus is the actual ancestor, of course, but only that the vertebrate ancestors were probably similar to amphioxus. Drs. Morris and Gish both discuss the transition from invertebrates to vertebrates, but they are both so concerned with demonstrating the absence of fossil forms that they have nothing to say about living animals. Perhaps they would dismiss amphioxus as being merely an unusual vertebrate and thus maintain that we still had not found a bridge across the invertebrate-vertebrate gap. Aside from the fact that it is *not* a vertebrate as it lacks a vertebral column, this is a reasonable argument in itself, except that amphioxus is not the only primitive chordate animal.

Sea squirts look nothing at all like vertebrates, not even to the most imaginative mind. They are sac-like creatures often about the size of one's thumb, and if you squeeze them you can squirt water from their two spouts. You can often spot them at low tide, attached to rocks. Many of them look so uninteresting that they would easily be overlooked, but others, mostly the ones that live together as colonies, are brightly colored and look most intriguing. They are filter feeders, like amphioxus, and the bulk of their structure comprises an elongate bag, the pharynx, which is perforated by numerous slits. These are called gill slits, but it requires some stretch of the imagination to compare them with the gill slits of a fish. Have they got any convincing chordate features at all? None. Here we have a regular-looking invertebrate. However, the larva of the sea squirt looks just like a small tadpole, and possesses most of the features that we saw in amphioxus: a notochord, a hollow dorsal nerve cord, and a pharynx, usually with one pair of gill slits. Here is persuasive evidence for the invertebrate-vertebrate connection.

After several days of free swimming, the tadpole settles on the bottom and changes into a sea squirt, never to roam again. There is nothing unusual about this alternation between a free-swimming larva and a sedentary adult phase; the same happens in other organisms. The barnacle, for example, spends its adult life clamped to rocks along the seashore, but its larva is a small, shrimp-like creature which spends its time drifting along with the rest of the plankton in the upper layers of the sea. In fact, a large percentage of planktonic animals are the larvae of more sedentary parents, and the purpose they serve is dispersal.

We are used to seeing a close resemblance between young and adult individuals; small children, kittens, and puppies look like scaled-down versions of their parents, but this is often not the case among invertebrate animals. Butterflies no more look like caterpillars than barnacles look like their shrimpy larvae. If some relatively small changes occurred during development to prevent a larva from developing into the adult form, a major evolutionary change would be effected. We can imagine that such a change may have led to the origin of the first chordates, and if this seems to be stretching credibility, consider the alternation in body form that we see within the coelenterates.

The coelenterates have always been one of my favorite groups of invertebrate animals, mainly because they look so attractive. What could be more beautiful than a brilliantly colored anemone, tentacles surging with the tide, or a jellyfish, its transparent bell pulsating rhythmically, or a coral at the edge of a reef. There would seem a world of difference between a jellyfish and a coral, but the connection between them is clearly shown by their life histories. One of the types of coelenterates which we have not mentioned is the hydroid. Many of these look like small plants, and they are often misidentified as such when students first see them. They are usually found growing on seaweed, and when viewed under a microscope they are seen to comprise a number of tentacled structures which look like flowers. These are the polyps, or the hydranths, and serve to gather food. Sometimes a second type of structure can be seen which bears small buds. These eventually become small jellyfish, called medusoids, which break free and swim off. Some of the medusoids are female and bear eggs; others, the males, shed sperms into the sea. The fertilized egg develops into a ciliated larva which, after a brief period of free swimming, settles on the bottom and develops into a new colony of polyps. There is therefore an alternation between the polyp phase and the medusoid phase.

In the jellyfishes proper the medusoid phase is the predominant one, and the polyp phase occupies but a relatively short part of the life cycle, serving only to bud off medusoids. Somewhat the reverse is true for the anemones, which have only a polyp phase, and no medusoid phase at all. Corals, which are essentially anemones that secrete a hard skeleton around themselves, similarly lack the medusoid phase. Which came first, jellyfish or anemones? For our purposes the answer is not important. What is important is that these two types of animals are interconnected — through the hydroid phase — and either one could have evolved from the other by a modification in the life cycle.

This brief survey of living organisms has shown that the creationists are wrong when they say that there are no connections between the major groups of organisms. We have not been able to document connections between *all* major groups, but this is partly because they are not all interrelated. To put this another way, it is likely that multicellular organisms evolved more than once, and that some groups, for example the sponges, are probably side branches that did not lead anywhere.

The fact that we cannot draw a firm line between plants and animals, or between unicellular organisms and multicellular ones, is difficult to reconcile with the creation model. Taken with the evidence for a link between the two major invertebrate groups (insects and worms), a link that is also documented by fossils, and between invertebrates and vertebrates, we have an overwhelming case for evolution. And we have not yet finished, for we have still to consider the fossil record.

Each of the major vertebrate transitions: fishes to amphibians, amphibians to reptiles, reptiles to birds, and reptiles to mammals, is documented by fossils, but, because of the vagaries of preservation, some transitions are more fully documented than others. We will be looking at these transitions in Chapters 10 and 11. Before we can discuss this fossil evidence we need to know something about the nature of fossils, and of the rocks in which they are found. These subjects are treated in the next two chapters.

SEVEN   *Rocks and Time*

 TWO MAJOR ISSUES will be discussed in this chapter: how rocks were formed and how old they are. According to the creationists, rocks were formed catastrophically, and this was accomplished in a relatively short time. This period of rapid formation dates back to the Noachian flood, and the maximum age of the rocks, they say, is around ten thousand years. Evolutionists believe that fossil-bearing rocks were formed under steady-state conditions, that the process is going on today just as it has in the past, and that the oldest rocks date back almost four billion years.

We call fossil-bearing rocks sedimentary because of the way they are formed. Fossils are sometimes found in other rock types, too, but this is unusual, so we will not consider them. Sedimentary rocks are formed by the accumulation and consolidation of fine particles, as when sediments settling on a lake bottom eventually become a mudstone. Sedimentation goes hand-in-hand with erosion. The sea cliffs that crumble under the pounding of the waves are carried out to sea, where they settle to the bottom, eventually forming new sedimentary rock. But the creationists tell us that there is no proof that this process occurs. According to Dr. Morris: "The sediments at the bottom of the ocean are still soft sediments, not solid rock."

Of course he is right. The sediments at the bottom of the ocean *are* soft, for a depth of tens of meters, but they grade into firmer material, and this, in turn, grades into solid rock. I think that Dr. Morris's misconception stems from the fact that scientists seldom collect samples that show the transition from soft ooze to firm rock, and the reason is largely a matter of technology and costs.

People interested in the soft upper layer, palynologists for the most part (people who investigate past climates and floras by studying fossil pollen grains), collect their samples using a coring device. This is essentially a hollow metal tube, about 8 cm in diameter and usually not more than about 10 meters long. Lowered to the sea or lake bottom and allowed to free-fall for the last few meters, the tube penetrates into the ooze and a sample can be collected in the same way that an apple-corer removes the core of an apple. The core sample is extruded from the tube using a plunger which is rammed down the length of the corer. This technique only works on fairly soft mud, and as soon as the bottom of the corer encounters anything very firm it stops going down any further. The top end of the sample is very squishy, but it becomes firmer towards the bottom, which has the consistency of soft clay. By contrast, people interested in the hard lower layers, usually economic geologists searching for oil, ignore the soft layer and drill their way down to the deep sediments; time is money, and oil exploration is an expensive business. Their core samples are therefore of solid rock.

People interested in how sediments accumulate and how they become consolidated into solid rocks, sedimentologists as they are called, are unlikely to be able to afford to charter a drilling rig, and therefore seldom get the chance to study complete cores of any great length. But this changed on August 11, 1968, when the drilling ship *Glomar Challenger* began her first voyage for the United States Deep Sea Drilling Project. This multi-million-dollar project, involving a team of several hundred scientists, has undertaken to collect deep-core samples from oceans around the world.

The preliminary results of this exciting program appear in a series of volumes which are published soon after completion of each voyage. Detailed descriptions of several hundred cores have now been given, and these contain a wealth of information on the process of rock formation. These cores show impressive sequences from soft and unconsolidated oozes at the top, into progressively firmer sediments beneath. Take, for example, the core sample taken at site number 94 in the Gulf of Mexico in March 1970. With a length of 660 m, the core grades from a soft ooze at the top into progressively chalkier material beneath, which becomes consolidated into soft chalk at a depth of 250 m. By a depth of about 440 m, this has become consolidated into a hard chalk. Another core, taken two weeks earlier at site 92, which was a little closer to land, changed from a silty clay at the top to a mudstone at a depth of about 260 m. Here is proof that creationists deny exists: soft sediments on the ocean floor *do* become consolidated into solid rocks.

Let us now take a closer look at an example of rocks in the making, this one from a little closer to home. To do this we will take a short excursion to a small lake on the outskirts of Toronto. Crawford Lake, nestled among trees on the edge of the Niagara Escarpment, is about an hour's drive from Toronto. It is not noted for its fishing, but this lack is more than compensated for by the interesting mud that accumulates on its bottom. During the warm months of the year, microscopic plants (phytoplankton) teem in the surface layers, and they seem to extract so much carbon dioxide from the water that dissolved lime is precipitated from the water as chalky particles. These particles sink to the bottom and are deposited as a white layer. The winter sediments, in contrast, are dark in color, comprising mainly organic matter. Because the lake has a relatively small surface area compared with its volume, it is poorly oxygenated. Consequently there are no bottom-living organisms to disturb the mud, which therefore retains its laminated structure. A slice taken through the mud would therefore have a striped appearance, each light band corresponding to summer, each dark band to winter. Undisturbed samples of the mud can be collected using an ingenious technique which has been extensively used by Dr. J. H. McAndrews and his colleagues in the Department of Geobotany at the Royal Ontario Museum. A 2 m metal pipe, some 8 cm in diameter and sealed at the bottom with a lead sinker, is filled with dry ice, then capped with a valve. The valve lets the released carbon dioxide gas out without letting the water in. The pipe is lowered to the bottom on a rope where it sinks vertically into the soft mud. The intense cold produced by the dry ice freezes the mud to the tube, up to a thickness of about 1 cm. After about twenty minutes the pipe is pulled free by a sharp tug on the rope and the core sample is hauled to the surface.

Because each light-dark couplet, or varve, corresponds to a year, it is possible to count back through time, starting from

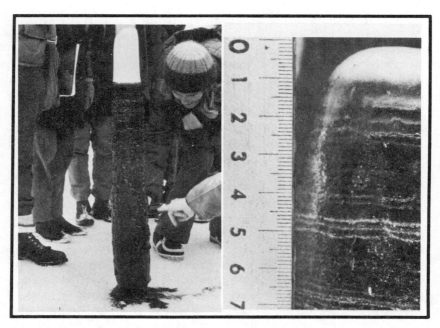

A core sample just collected from a lake bottom. Close-up of the core shows light and dark layers corresponding to summer and winter.

the layer at the top of the pipe, which corresponds to the present. The core, which is just under a meter long, dates back about 2000 years. Leaves and pieces of twig are often found sticking out of a particular layer, and the actual year when these became embedded in the sediments can be determined. Very occasionally a small fish skeleton has been found and this, like the leaves and twigs, is a fossil in the making. Pollen grains, which can be identified with particular species of plants, are an integral part of the sediments, and their study gives a clear picture of the floral changes that have occurred in the area over the years; they too are fossils. The appearance of maize pollen around the year 1300 marked the arrival of Indians (an Iroquois village was subsequently excavated in the woods, complete with charred maize kernels), while the clearing of the forests by European settlers is marked by the appearance of ragweed pollen.

If we compare the thickness of the layers at the top of the core with those further down, we see that they become thinner. This is because the muddy particles become more closely packed under the pressure of the layers accumulating above them. The pressure at the bottom of a 1 m layer of mud is obviously quite small (in the order of $1/5$ of an atmosphere — the pressure a swimmer experiences at the bottom of a swimming-pool), so the bottom layer would have to undergo considerably more compression before the water was squeezed out of it and it became a mudstone. Such consolidation would reduce it to a fraction of its present thickness. In time, then, our 2000 years of Ontario history would be compressed into a thickness of far less than one meter.

Without even making allowances for compression, if our one meter of Crawford Lake mud represents 2000 years, how many years is represented by, say, the 2000 m of sedimentary rocks forming the Grand Canyon? Obviously a good deal more time than the 10,000 or 20,000 years that creationists believe to be the age of the earth! And this does not take into account the fact that the Grand Canyon represents just a fraction of the total thickness of sedimentary rocks. On the evidence of sedimentary rates alone, it is quite clear that the earth is considerably older than creationists would have us believe it is. Radiometric dating sets the age at about 4 billion years.

To understand how radiometric dating works, we have to know a little about radioactivity. Certain elements, including uranium and thorium, are unstable, and their atoms undergo spontaneous disintegration, or decay. During this process they throw out high-energy particles and waves, collectively known as radiation, and they are therefore said to be radioactive. What is left is a smaller atom of another element, and this is usually unstable too. The process of decay and emission of radiation carries on until the resulting product is a stable element. The eventual end-product of the decay of uranium is lead, but the process takes a long time. We obviously cannot follow the disintegration of a particular uranium atom all the way to lead and see how much time it takes, but what we can do, based on actual observations, is to estimate how long it would take for half of the atoms in a sample of uranium to decay to lead. This quantity is termed the half-life. Many elements have two or more forms, called isotopes, differing from each other only in the weight of their atoms. Uranium has two important isotopes, written in chemical shorthand as U^{235} and U^{238}. The least stable isotope of uranium, U^{235}, has a half-life of 713 million years, compared with 4500 million years for U^{238}. This means that if we had 1 kg of U^{235}, in 713 million years we would have only

500 gm, and in another 713 million years this would be reduced to 250 gm, and so on. We would have correspondingly increasing amounts of lead.

If we can measure the amount of uranium in a rock sample, and the amount of lead that has accumulated from its decay, we can calculate the time it has taken for this decay to occur, given the half-life of the uranium in the rock. Creationists concede that we can accurately measure the amounts of uranium and lead in rock samples, but they argue that there is no way of knowing what the *initial* amounts of uranium and lead were. Obviously if there were already some lead present at time zero, and we did not know how much there was, our results for the age of the rock would be meaningless. While this seems a formidable problem, in actual fact it is not. Lead, whose chemical symbol is Pb, has four isotopes: Pb^{204}, Pb^{206}, Pb^{207}, and Pb^{208}. The isotopes Pb^{206} and Pb^{207} are the final decay products of U^{238} and U^{235} respectively. The isotope Pb^{208} is the final decay product of the element thorium (chemical symbol Th), which often occurs with uranium. The isotope Pb^{204}, however, is not formed by radioactive decay, and is present in all lead samples at an essentially constant concentration of about 1.5 percent. By measuring the amount of Pb^{204} in the rock sample, which, of course, is the same as it was initially because it is not generated by the radioactive decay, we can calculate the initial amounts of the other lead isotopes. Furthermore, since the other three lead isotopes are the decay products of three different radioactive sources, each with their own separate half-lives, three separate age estimates can be calculated. This allows cross-checking of the age determinations.

Creationists often question the constancy of half-lives, arguing that we cannot assume that they have remained unchanged. This, of course, is true; we cannot be absolutely sure that half-lives, day lengths, or anything else will remain constant. All that we can do is make predictions based upon previous observations. I am writing this sentence at noon, and it is light outside. It was also light yesterday at noon, and that is the way it has been for as long as I can remember. It *might* be dark tomorrow at noon, but I have no reason to suspect that it will be. The rate at which particular radioactive atoms disintegrate has been measured by direct observation and has been found to be constant. It is not changed by temperature, pressure, or chemical reaction, so it is difficult to visualize how it could be changed. When we make the assumption that half-lives are constant, then, we are making a predictability statement that might prove to be wrong, but we have no reason to believe that it will. Actually, it is rather ironic that creationists should question the constancy of half-lives when they tell us that, according to their creation model, the "basic nature of matter" is constant. They do tend to want to keep their cake and eat it too.

What about the criticism that radiometric dates vary so widely as to be useless? My response is to point out that there are good radiometric labs and bad ones, just as there are good doctors and bad ones. Radiometric dating requires precise analyses under the most carefully controlled conditions. When measuring quantities of lead in parts per million, all extraneous lead has to be eliminated. The air we breathe contains lead, from automobile emissions, and it is therefore necessary to exclude this from the apparatus. The fact that a number of laboratories throughout the world have produced hundreds of radiometric dates throughout the geological column that closely correspond to one another bears testimony to the reliability of the methodology.

Another criticism of radiometric dating that creationists make is that scientists first decide the approximate age of their sample, then choose the appropriate radioactive series (uranium-lead, potassium-argon, Carbon [14], or whatever) to give the right age. For example, if they had a piece of mammoth tusk to date, they would choose the Carbon [14] method, because the relatively short half-life of Carbon [14] (5700 years) gives ages in thousands rather than in millions of years, whereas a piece of dinosaur bone might be dated using the potassium-argon method, which gives ages in tens of millions of years. This is comparable to criticizing a chef for using an egg-timer when boiling eggs instead of a sundial.

One point that should be made here is that we cannot radiometrically date any old piece of rock or bone. We can only analyze materials that contain radioactive elements. Furthermore, we can only analyze materials that were formed at the same time as the specimen we are trying to date. For example, if we wanted to date a particular level in a formation of sedimentary rocks, it would be no use dating a sand grain or a pebble incorporated into the sediments from elsewhere. If we did, we would only get a date for that older rock formation that became eroded and incorporated in the new sedimentary rock. We therefore have to restrict ourselves to materials within the sediments that we know were formed at the same time as the sediments enclosing them, such as volcanic ash and lava.

Creationists, having been so critical of radiometric dating, would surely have some rigorous method for arriving at their young age for the earth. How do they do it? They use such infallible methods as assessing how long it has taken for our 3.5 billion world population to grow from the original nuclear family of Adam and Eve! Dr. Morris calculated that if Adam and Eve, and subsequent couples arising from their union, had only between two and three offspring, the present world population could have been reached in only one hundred generations. Assuming a generation time of forty years, which is very reasonable, this represents only four thousand years. Applying similar logic to the evolutionists' 1-million-year estimate for the history of man (which is probably very conservative), he arrives at a number of such astronomic size that it exceeds the number of electrons in the universe. Impressive stuff!

There is nothing wrong with Dr. Morris's arithmetic, but his biology is off base. Now flies, as we all know, lay thousands of eggs and their generation time is measured in weeks, not years. We should therefore expect to be over our heads in the things by now, but we are not, and the same is true for other animals. The reason is that populations are kept in close check by competition, especially for food; by disease; by adverse climatic conditions; and, it has been suggested, by internal regulatory mechanisms. Population numbers do fluctuate from year to year, but over the long term they remain essentially constant, and this has been observed for a wide variety of animals, both vertebrate and invertebrate. These various controls do get out of order at times, and then we see population explosions — swarms of locusts, flocks of lemmings, and the like — but over the long term, animal populations remain fairly constant. The notable exception is our own species, and the reason we are experiencing a population explosion, a fairly recent phenomenon, is that we have learned to eliminate competition and to overcome disease. We do not share our crops with any of the other species, not with rabbits (we all but wiped them out with myxomatosis in the fifties), or with insects (our new pesticides are notoriously deadly), and we do not allow other plants (weeds) to compete with them. This, of course, is true only in the "developed and developing world"; primitive cultures are nowhere near as successful in competing with nature. This reflects in their small population sizes and also in the numbers of tribes that have become extinct. While many of these extinctions have probably been caused by interference from "civilized" peoples, the fact that they did occur reflects such tribes' precarious tenure on the earth. Consider, for example, the Fuegians, living in isolation on the bleak island of Tierra del Fuego, off the southern tip of Argentina. When Darwin visited the island in 1832 they numbered in the hundreds, but by the end of that century they were all but extinct.

The creationists, resourceful people that they are, have several more "scientific proofs" for a young earth, but none stand up to close scrutiny. Take, for instance, their argument about the fall of meteoric dust on the earth. Citing the work of Hans Pettersson, which appeared in the respected *Scientific American* in 1960, Dr. Morris tells us that an incredible 14 million tons (as these are estimated figures, we can read metric tonnes — there is only about a ten percent difference between metric and English tons) falls each year, derived from disintegrating meteors. He then calculates that over a 5-billion-year period this would have accumulated into a 55 m (182 ft.) thick layer over the entire globe! Of course there is no such layer of dust on the earth, and Dr. Morris rejects any suggesting that this lack could be accounted for by mixing processes. This would appear to be a formidable problem for evolutionists to overcome, but it is not, and we shall see that Dr. Morris was both selective in reporting Dr. Pettersson's work, and confused.

First, in the same article quoted by Dr. Morris, Dr. Pettersson gave an estimated range of 5–14 million tons for the rate of fall, preferring the lower estimate: "To be on the safe side, especially in view of the uncertainty as to how long it takes meteoritic dust to descend, I am inclined to find five million tons per year plausible. The five-million-ton estimate also squares nicely with the nickel content of deep-ocean sediments." Dr. Pettersson also makes the point that the rate of accumulation appears to be higher now than it has been during the past ten or fifteen million years.

Five million tonnes falling each year still sounds like an enormous amount, but, spread evenly over the globe, this is only about 10 kg per square kilometer — probably a shoeboxful (this material has a high density, comparable to that of iron). Imagine sprinkling this quantity of dust over two hundred football fields: it would be very difficult to detect. Dr. Pettersson tells us that the dust is so rare in ocean sediments that one has to select areas with a very slow sedimentation rate (one or two millimeters in one thousand years) to stand any chance of collecting a sample. Dr. Morris's contention that mixing cannot account for the "missing" 55 m of meteorite dust is as absurd as his referring to it as a discrete layer.

The creationists also have other pieces of "evidence" for a youthful earth with which to confound us. The earth's magnetic field, as so many creationists have told me, is changing. If we extrapolate this rate of change back in time, beyond about ten thousand years, the field would be as powerful as that of a magnetic star. This sounds so unlikely to the creationists that they take it as evidence that the earth could not possibly be more than ten thousand years old. What they have overlooked, however, is that the earth's magnetic field undergoes cyclical changes and we have evidence of this in certain rocks, in the form of magnetic stripes. It was the matching up of these stripes, from one continent to another, that led to the general accept- ance of the theory of continental drift — the idea that conti- nents were once in contact and have slowly drifted apart. This is a theory that the creationists find convenient to believe in when it appears to support their arguments.

I was driving back to Toronto one winter's night with one of my good creationist friends when we got to talking about Noah and the flood. "He would have to have dropped all the animals off to their appropriate continents after the flood, wouldn't he?" I asked. "Over to Africa with the lions and elephants, across to India with the tigers, down to Australia with the kangaroos — like dropping the kids off after a hockey match."

"Oh no," was the reply. "Continental drift would take care of all that."

Now if creationists are going to accept continental drift, they have to accept radiometric dating, because the matching up of the edges of continents was based upon radiometric dates. But that is not all.

"Hang on," said I. "If you believe that the earth is only about ten thousand years old, there has not been enough time for the continents to drift apart. They are only moving at about 5 cm per year."

"Ah," said my friend with some conviction, "the rate of drift has slowed down since those days."

You just cannot win!

Aside from the overwhelming geological and radiometric evidence for an old earth, there is incontrovertible evidence from astronomical measurements. Because of the enormous dis- tances involved, astronomical measurements are usually ex- pressed in terms of light-years: the distance that light travels in one year. Considering that light travels at a velocity of 340 000 km (186,000 miles) per second, this is obviously a con- siderable distance. The nearest star, Alpha Centauri, is four light-years away. This means that when we look up at the night sky and see this star, we are actually seeing it as it was four years ago, because it takes that long for the light to reach us (it takes about eight minutes for the light to reach us from the sun). Consider the following distances: the width of our galaxy is 100,000 light-years, the nearest galaxies are within 2.5 million light-years, many thousands of galaxies are within 50 million light-years, and the most distant galaxies are farther than 5 billion light-years away. This means that when we look at these distant bodies through our powerful telescopes we are seeing them as they looked millions of years ago. From this it follows that they must be millions of years old. How can creationists possibly rationalize this with their belief that the universe is only about ten thousand years old? Since nearly all astronomical measurements are based on the method of triangulation, which has been used by surveyors since the time of ancient Greeks, they are difficult to refute. Furthermore, we have been able to confirm the accuracy of our measurements using space probes. If our assumptions were wrong, or our methods inaccurate, Viking Lander would never have found Mars, let alone landed!

During a recent encounter I had with Dr. Gish, he was careless enough to make some reference to astronomical distances in terms of millions of light-years. When I asked him how he could rationalize this statement with his belief in a ten-thousand-year-old universe, he was unable to give a satisfactory reply.

There is no question that the earth is very old.

EIGHT

Fossils: How They Are Formed and What They Can Tell Us

THE PURPOSE OF THIS CHAPTER is threefold: to show that fossils are formed under steady-state conditions rather than under catastrophic conditions as the creationists claim; to show that fossils represent a small fraction of all organisms that have ever lived; and to acknowledge the limitations of fossil evidence.

Have you ever wondered why we do not find more dead animals lying around than we do? City authorities, of course, are fairly efficient at keeping our streets clean, but dead animals are almost as uncommon in the countryside, too. Just consider how many thousands of birds perish each year during our cold Canadian winters, yet dead birds are seldom seen. The reason is that their corpses have been devoured by scavengers.

Field observations on carcasses confirm how rapidly scavenging animals go about their black business. A group of scientists studying the orangutan in a Borneo rain forest found the remains of an old male ape. He had died within hours but wild pigs had already gone to work. The entire contents of the body cavity had been ripped out and devoured, many ribs were broken, and much flesh had been stripped from the bones. Even as they examined the carcass the pigs returned and carried on their work, wrenching limbs from their sockets and carrying them away. The inhabitants of forests and grasslands are swift at destroying their dead and that is why our own evolutionary history, which was made in these very habitats, is so incomplete.

The only chance of a skeleton's being preserved is if the corpse escapes the attention of scavengers, and the most likely way that this can happen is through burial. Burial might be through dust or sand storm, through silting up on a lake, river, or sea bed, or through entombment in a cave. These processes, to my mind, seem fairly ordinary, but I do not have the vivid imagination of a creationist. Dr. Morris insists that "it should be obvious that the actual formation of potential fossils in the first place . . . requires rapid and compact burial of the organisms concerned, and this requires catastrophism." To support his point he cites several references to fossil localities where bones are especially rich, localities variously referred to as "grave-yards" and "boneyards." Their existence, he alleges, proves that fossils are formed only under catastrophic conditions.

Fossil-rich localities like these, sometimes with bones piled one on top of the other, do exist, but they are the exception rather than the rule. Armchair paleontologists write about fossils being found "more often than not in the form of fossil graveyards containing large numbers of fossils," but real paleontologists know how many hours and days of foot-slogging and patient searching lie between one fossil find and the next. It must also be emphasized that such occurrences are always local both geographically and historically; there is no evidence whatsoever of any fossil graveyard extending across continents.

I do not deny that certain burial sites do suggest that some sort of local catastrophe has occurred. The famous Jurassic quarries of Holzmaden, in southern Germany, come to mind, where particularly well preserved skeletons of ichthyosaurs (marine reptiles superficially like dolphins) are found in large numbers at some horizons, suggesting mass burial. However, I believe that most fossils have probably been formed under steady-state conditions similar to those pertaining today, and Dr. Morris's challenge to "try to think of places where fossils are being formed today by uniformitarian processes" is not very challenging at all.

We have already seen examples of pollen, leaves, and the occasional fish skeleton as fossils in the making on a lake bed in southern Ontario. For scuba divers, dead fish lying on the ooze at the bottom are not an uncommon sight, especially in poorly oxygenated lakes where there are no bottom-feeding scavengers. They can be seen in various stages of burial, all of them fossils in the making. When we discussed deep-sea cores we could have pointed out that the soft sediments at the top, the rocks in the making, are invariably crammed full with dead marine organisms, especially foraminiferans (minute shelled animals). Similarly, the hard rocks below, with which the soft sediments merge, are replete with fossils. Indeed, it is these fossils that give geologists clues to the mineral potentials of the rocks they are drilling through. For some other examples of fossils in the making we will visit some caves and a swamp on the other side of the world.

While in New Zealand a few years ago, I spent a little time looking for moa bones. Moas are large, flightless birds, related to the emu, which became extinct some time during the period of colonization by Polynesians, that is, within the last six hundred years. Skeletal remains of moas, some with skin, dried meat, tendons, and even a feather or two still attached, are quite common in many parts of the country. The first locality I visited to do some exploring in caves was set in rolling chalk country just outside Wellington. These were not the great underground caverns that come to mind when we think of caves, but rock fissures and sinkholes, most of them just about large enough to clamber down into. Because their openings are often easily overlooked, they form natural traps, and the soil which accumulates at the bottom is a veritable graveyard. Aside from moa bones, I found remnants of kiwis (still living in New Zealand), *Sphenodon* (the lizard-like animal mentioned in Chapter 3 as an example of a "living fossil"), several ground-dwelling birds, and, in one cave, a particularly smelly sheep. I am sure that the sheep regarded his neck-breaking plunge to the bottom of the cave as catastrophic, but I doubt whether it was accompanied by bolts of lightning, tempest, and all the usual trappings of a good old-fashioned catastrophe.

A few miles outside the beautiful city of Christchurch, on South Island's Canterbury Plain, is a low-lying, swampy area called Pyramid Valley. Moa bones were discovered there in 1937 and since that time hundreds of skeletons, many of them complete and all of them extremely well preserved, have been collected. It seems that these great birds became mired in the mud, and many of the skeletons were found in an upright standing position, just where they had become trapped. Being mired in a swamp was certainly serious from the moa's point of view, but this could not be described as a New Zealand catastrophe, far less a global one.

There is obviously nothing supernatural about fossilization. That is not to say that we thoroughly understand the process, because we do not, but we do not have to invoke catastrophes to explain it. Nor do we have to go far afield to see fossils in the making. All we have to do is to know where to look, and how to see. The problem is that creationists have never taken the trouble to look for fossils in the making, and they are the very ones who are telling us that they just do not exist.

We will now consider what the chances are of a body's being preserved and subsequently unearthed as a fossil. In this way we will be able to gauge the completeness of the fossil record.

Escape from scavengers, as we have seen, is a prerequisite for fossilization, and this is most likely to occur through rapid burial, or through coming to rest in a cave, or on a lake bed which is free from scavengers, or in some similar location. The chances of this happening are quite low, especially for large bodies. Even when a body has escaped the attention of scavengers, it is quite possible that it will be destroyed by bacteria, or by the activities of acids in the sediments. The chances of preservation, then, must be small, but there are still more weeding-out processes to come. The sediments in which the fossil is entombed must survive erosion and geological upheaval. The fossil then has to be exposed in the rock by the weathering action of the elements, in a locality accessible to paleontologists. Think how many tonnes of fossils are weathering out of the rocks at this very moment, never to be seen, and certainly never to be collected. They may be weathering out of a mountainside, a remote seashore, a river bed, or some other inaccessible region not visited by man. There must be far more fossils going to dust on the wind than are ever collected. There is no question, then, that the fossil record represents the merest fraction of organisms that once inhabited this earth. That is why the fossil record sometimes gives us a big surprise, as when the remains of gigantic flying reptiles (pterosaurs), with their incredible 15 m wingspan, were recently found in Texas, and when a rare nursery of baby dinosaurs was found in Montana in 1976.

Darwin was perfectly correct, then, in blaming some of his problems on the incompleteness of the fossil record, but most of his problems arose because he was trying to use it to document the origins of new species. As we will now see, the fossil record lacks the fine resolution needed to chronicle such small-scale evolutionary changes.

We paleontologists are sometimes very naughty; we sometimes

try to infer too much from fossils. When a paleontologist describes how an animal walked, what it ate, or how fast it ran, all on the evidence of a few bones, he is not just skating on thin ice, he is walking on water. There is only so much information that can be gleaned from studying old bones and teeth, and it is important to be aware of this limitation. At best, a paleontologist has a complete skeleton to work with, and usually he has much less. While this skeleton can answer a good many questions, many more remain unanswered. We can see whether a fossil has the sharp teeth of a carnivore or the grinding teeth of a herbivore, but we cannot tell whether it had the voracious appetite of a warm-blooded animal or the modest appetite associated with cold-bloodedness. We can tell how big an animal was, but not its color. We may surmise that it walked on all fours, but we cannot say how fast. We might conclude it was an adult, but we almost never know its sex. We can probably identify it as belonging to one particular species, but we cannot be sure that individuals of that species were actually able to freely interbreed when they came into contact. In short, we can say nothing more about the fossil than we could say about any living animal, given only its skeleton to work with.

This does not mean that skeletal features cannot be used to identify living species, because they usually can. Mammals, for example, are usually identified by their skulls and teeth, molluscs by their shells. The point we are making is that species cannot always be distinguished on skeletal features. Furthermore, since we can never know whether the species we identify in the rocks are true biological species, we cannot expect to be able to recognize the origin of a new species in the fossil record. Therefore, if we had a fossil sequence where a new biological species was actually arising from another species, we would probably be unable to recognize it. The fossil record lacks the fine resolution to record small-scale evolutionary change, and this is why it was such a disappointment to Darwin. What does it offer at the other end of the scale, in the documentation of major evolutionary change?

Even to the untrained eye the differences between the fossil remains of, say, a crocodile and a horse, or an ammonite and a coral, are absolute, and we rarely have any difficulty in assigning a fossil to its appropriate major category. Consequently, the fossil record, as we will see in Chapters 9–14, provides us with a good record of many of the major events in the history of life. The record is incomplete, though, as we have said earlier, and this is partly because of the small chance of a body's being preserved, and partly because hard parts — bone, teeth, shells, seeds, wood and the like — are usually all that we have to work with. Soft parts, such as skin impressions of dinosaurs, and soft-bodied animals like jellyfish are sometimes preserved, and in some localities may be common, but they give us only brief glimpses of evolutionary histories. Obviously we have no record of the origin of life, and little or no evolutionary history of the soft-bodied organisms. It is hardly surprising, then, that we have so many gaps in the evolutionary history of life, gaps in such key areas as the origin of the multicellular organisms, the origin of the vertebrates, not to mention the origins of most invertebrate groups. The creationists, of course, just love to draw attention to these gaps, which they score as points against evolution. We saw in Chapter 6, though, that their case is without foundation, because they have ignored vital evidence from the living world.

Another implication of the incomplete fossil record is that it gives only part of the evolutionary history of a given group. This is simply because so many major evolutionary changes have taken place in the soft parts of the body, and in other non-fossilized features such as behavior, physiology (body mechanisms like digestion and excretion of waste products), and reproduction. The major advances in the evolutionary history of the mammals, for example, were probably the attainment of warm-bloodedness, the nourishment of the embryo through the placenta, the development of mammary and sweat glands, and extended parental care of the young, none of which appear in a fossil record. While changes in the skeleton and the teeth have often accompanied major changes in the soft anatomy, this was not always the case, and such changes were not necessarily coincident. Mammals, for example, probably evolved a secondary palate (roof to the mouth) before they evolved a constant body temperature, while birds evolved feathers before losing their long reptilian tail (see Chapter 9).

When we see gaps within a segment of the fossil record, we should remember that these gaps may not necessarily be as significant as they appear, because they may be bridged by features that are not preserved. For example, if we compared skeletons of the two living groups of whales, the toothed whales (including dolphins) and the baleen whales (including the gray whale), we would see many differences between them, especially in the skulls. But when we compare other features, such as the tail fluke, which is not supported by any

skeletal structure, the complex brain, and the breathing apparatus, not to mention their complex behavior, which includes
vocalization and much more besides, we realize that the two
groups have a great deal in common and the gap between them
is narrowed.

So far in this chapter we have made the following points:

- Fossils can be, and probably usually are, formed under
  steady-state conditions. We can actually see the process going
  on around us today.
- The chances of preservation and subsequent discovery of a
  fossil are remote. The fossil record is therefore very incomplete.
- The fossil record is unsuitable for documenting the origins
  of new species, but illustrates many of the major evolutionary changes that have occurred in the history of life.
- The fossil record is essentially an evolutionary history of the
  changes that have occurred in the skeleton and the other hard
  parts of the organisms. Evolutionary histories of the soft anatomy, behavior, physiology, and reproduction, and also of soft-
  bodied animals, are therefore either absent or very incomplete.
- Because we see almost nothing of the evolutionary history
  of soft body parts, and their functions, the gaps in the evolutionary records may be an exaggeration of the true situation.

Before going on to the next chapter, we have one more aspect
of fossils to consider, which relates to our system of classification rather than to the fossils themselves. Creationists argue
(erroneously, as we shall see in Chapters 10–14) that there are
no fossils that are intermediate between one major group and
another. The main reason they make this argument is their
unfamiliarity with fossils, but a small part of the blame has to
be borne by paleontologists. As I said earlier, paleontologists
seldom have difficulty distinguishing between major groups,
but they often have problems distinguishing between fossils
that lie close to the border between two groups. Modern
mammals can easily be distinguished from modern reptiles by
their skeletal anatomy. As we shall see in Chapter 11, if we
compared a lion with a crocodile we would find, among many
other features, that the crocodile has simple, conical teeth and
a small braincase, while the lion has complex teeth, with cusps,
and a large braincase. There are some fossils, however, appro-
priately called mammal-like reptiles, that have a small cranium
but complex teeth with cusps. Indeed, these fossils have a
whole mixture of reptilian and mammalian features. How
should we classify them? Paleontologists are very tidy-minded

people who like to see organisms placed into their appropriate categories. In the case of mammals, paleontologists decided to base their ruling on the sort of jaw joint the beast has — if the joint is formed between the articular bone and the quadrate bone it is called a reptile, but if it is formed between the dentary and the squamosal bone it is classified as a mammal. As if to confound the system, a fossil was described in 1970 that had both sorts of joint! But paleontologists were not going to be beaten by a fossil that could not decide what it wanted to be, and responded by adding a second requisite, concerning the teeth. To be classified as a mammal a fossil has to have a jaw joint formed between the squamosal and the dentary bone *and* have accessory cusps on the main tooth cusps.

There are some lessons to be learned here. Our classification system is neither natural nor perfect, and if we assign fossils (or modern organisms) to one or other category on the basis of only one or two features, we are bound to make mistakes. More important still, our classification system fails to recognize the existence of transitional forms. Obviously, if a group is defined on one feature, or a complex of features (in our example, jaw joint plus teeth), transitional forms, by definition, can never occur. Suppose the fossil in question has the dentary-squamosal jaw joint *and* has accessory cusps? Obviously it is classified as a mammal. But if another fossil has only one of the features in the mammalian condition, it is classified as a reptile. I wonder how many intermediate forms, both fossil and living, we have hidden away in our classification system?

NINE   *Fossils and Time*

ACCORDING TO THE CREATIONISTS, the succes-
sional arrangement of the different types of
fossils in rocks of progressively younger age
(called the geological column) is an illusion,
based upon preconceived ideas of evolution.
They claim, instead, that fossil organisms lived
together contemporaneously, and that their burial in the sediments (from the Noachian flood) occurred over a period of
months rather than millions of years as claimed by evolutionists. We will show in this chapter that they are wrong on all
counts.

Devil's toenails, serpent's eggs, and thunderbolts; these
names have been applied to such common fossils as oysters,
sea urchins, and belemnites, and reflect the mysticism that was
once associated with fossils. Not until the seventeenth century
did people come to realize that fossils were the remains of
organisms that had once lived on the earth, and their importance in interpreting the remote history of our planet did not
follow until much later. So it was that the earliest geologists,
men like Lehmann, Füchsel, and Werner, working during the
eighteenth century, paid little heed to fossils. They preferred
to recognize the various rock formations on the features of the
rocks themselves. The Chalk, for example, which was easily

recognized, could be followed right across western Europe and was used to mark the upper limit of what were called the "Secondary" strata (the Mesozoic Era). The early importance of lithology (the study of the nature of rocks) is shown by the names that were used by the early geologists, names like Old Red Sandstone, Oxford Clay, Greensand, Portland Stone, and Lias (from the Gaelic *leac*, a flat stone), and are still in use today.

When attention was paid to fossils, it was found that certain kinds were restricted to certain rock layers. Giovanni Arduino (1739–95) coined the terms Primitive, Secondary, and Tertiary for the main groupings of rock formations, identifying them on their fossils. According to Arduino, the Primitive rocks were unfossiliferous, the Secondary rocks contained mostly marine fossils, and Tertiary rocks were highly fossiliferous. (The term Tertiary is still used today.) The principle which recognizes that certain fossils are restricted to certain geological horizons is called stratigraphy. The Englishman William Smith (1769–1839), whose job as a surveyor allowed him to spend a great deal of time examining rocks, used the principle of stratigraphy to trace particular rock layers across large areas of the country. He published his *Tabular View of the British Strata* in 1790, and completed a geological map of southern England in 1815. Georges Cuvier, the great French anatomist and anti-evolutionist (he soundly rejected fellow-countryman Lamarck's evolutionary hypothesis in a series of exchanges in the early years of the nineteenth century), had made similar observations during his studies of the geology of the Paris Basin. He was able to distinguish seven major formations that lay above the Chalk, and subdivisions within these formations, all based upon the index fossils they contained. What is more, he recognized that there was a definite succession in the types of fossils that were found as the geological column was ascended. He saw, for example, that reptiles predominated in the Secondary strata (Mesozoic Era), while Tertiary strata were characterized by mammals.

Darwin was only a baby when all this was going on, and it is quite untrue to say that the succession of fossils throughout geological time had anything to do with any preconceived ideas of evolution. The geological column was conceived by geologists as a direct result of their field observations. The creationists argue, however, that the geological column is based upon the assumption of evolution. Evolutionists, they say, therefore place the invertebrates lower in the geological col-

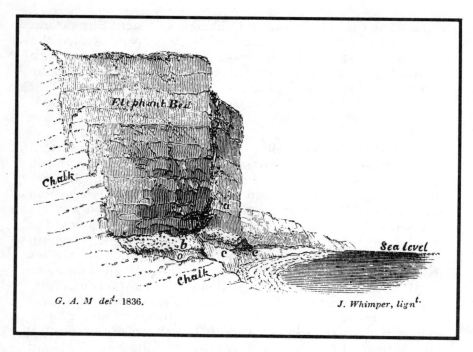

The occurrence of different types of fossils at different levels in the geological column was known to geologists long before Darwin's Origin of Species. *This engraving dates from 1836.*

umn than the vertebrates simply because of their belief that the former evolved before the latter. To be fair, though, I must point out that during a recent public forum Dr. Gish did concede the error when I confronted him with a geological map published in 1795.

The succession of different kinds of fossils throughout time demolishes the creationists' argument that all living organisms were created at the same time, obliging them to take desperate steps to defend themselves. What to do? What link in the fossil chain to try and put asunder? Of course — the Precambrian fossils.

The Cambrian Period, which began about 570 million years ago, is marked by the sudden appearance of a wide diversity of fossils, but before this, during the Precambrian Period, there is nothing. That, at least, is the way the story used to be told until the 1950s. Darwin himself was dismayed at the apparent suddenness with which life first appears in the fossil record. "There is another . . . difficulty which is much graver. I allude

Period	Time (millions of years)
	2
Tertiary	
	63
Cretaceous	
	135
Jurassic	
	180
Triassic	230
Permian	
	280
Carboniferous	
	345
Devonian	
	405
Silurian	
	425
Ordovician	
	500
Cambrian	
	570
Precambrian	

The geological time scale.

to the manner in which numbers of species . . . suddenly appear in the lowest known fossiliferous rocks. . . . before the lowest [fossiliferous] stratum was deposited, long periods elapsed . . . vast, yet quite unknown periods of time." His reference to vast periods of time was not overstating the case, because the Precambrian Period, which dates all the way back to the origin of the earth some 4.5 billion years ago, accounts for about ⅞ of the history of the earth. In geological terms, then, the appearance of life on earth did occur suddenly, and creationists have seized upon this like drowning men grabbing at straws. There is, however, some division in the ranks. Dr. Gish denies the existence of Precambrian fossils: *"Not a single, indisputable, multicellular fossil has ever been found in Precambrian rocks!"* (his italics)

Dr. Morris is a little more cautious, and makes the following points, which I have summarized to convey the general theme of his arguments.

1. All kingdoms and subkingdoms are represented from the Cambrian onward.
2. All animal phyla (major groups) are represented from the Cambrian onward.
3. All plant phyla are represented from the Triassic onward, except for bacteria, algae, and fungi (Precambrian onward), bryophytes and pteridophytes (Silurian onward), spermatophytes (Carboniferous onward), and diatoms (Jurassic onward).
4. All animal classes (next major group after phyla) are represented from the Cambrian onward, except for vertebrates and crinoids (Ordovician onward) and insects (Devonian onward).

Notwithstanding that Dr. Morris has some of his facts wrong, it is obvious that he does recognize that there is a sequential appearance of different organisms through the geological record. Not only is this incompatible with the creation account, but it corroborates the theory of evolution. How does Dr. Morris deal with this problem? Quite simply: he ignores it until the next chapter, where he explains it all away in terms of the different ways that organisms responded to the rising waters of Noah's flood, and the different rates at which their bodies sank. We refuted these imaginative arguments in Chapter 5.

One of the main reasons for the misconception that fossils appeared suddenly in rocks of Cambrian age, without antecedents, is due to the difficulty of recognizing the start of the Cambrian. There is no obvious rock feature that can be used to distinguish between the end of the Precambrian and the beginning of the Cambrian, and the most obvious feature of Cambrian rocks is the abundance of marine organisms: seaweed, jellyfish, various shellfish, worms, and trilobites (segmented animals with several pairs of jointed legs and an external skeleton). Paleontologists have therefore tended to use the appearance of fossils to identify the start of the Cambrian. As progressively older fossils were found, the base of the Cambrian got pushed further back, so that Precambrian fossils, by definition, could not exist. With improvements in radiometric dating it could be shown that fossil-bearing strata extended back over a considerable expanse of time, and much work has been done in the last twenty years or so to redefine the start of the Cambrian. The period started about 550 million years ago.

Are there any fossils in Precambrian rocks? There are no Precambrian trilobites, or shellfish, but large numbers of multicellular fossils have been found that are identified as jellyfishes and coelenterates (these are among the simplest of multicellular animals), together with some segmented worms (annelids), and an animal that looks something like a sea urchin. Precambrian fossils have been found at Cape Race, Newfoundland, in Leicestershire, England, and in Namibia, Russia, and South Australia. Radiometric dates for the fossils extend back to about 700 million years ago, but these are not the only, or the oldest, Precambrian fossils.

The quest for fossil evidence of unicellular organisms would appear to be a futile one. What, after all, would seem less likely to be preserved than a tiny blob of protoplasm? Yet however impossible it might seem, a large number of unicellular organisms have been preserved, and many of these have been found inside layered stony structures called stromatolites. Stromatolites were discovered at the beginning of the present century in Precambrian rocks of western North America. These short columns, whose appearance has been compared with a stack of pancakes, look like some sort of rock, but they were interpreted by their discoverer as having been formed by the action of colonial algae. Other scientists were skeptical, however, and stromatolites were generally considered to have been formed by non-biological processes. Then, in 1954, a scientist working at Harvard University reported that he had found some microscopic fossils inside a piece of stromatolite collected from the Gunflint Iron Formation in Ontario. The fossils were very small (in the order of 60 micrometers — a micrometer is $^1/_{1000}$ mm) and looked very much like modern blue-green algae.

As we saw in Chapter 6, these organisms, which lack a nucleus, are the most primitive of cells and are called prokaryotes.

Since this discovery, living stromatolites have been found on the coast of Western Australia, and microscopic examination confirms that they are formed by colonies of blue-green algae. These cells look strikingly similar to the fossil cells found inside fossil stromatolites.

While the preservation of unicellular organisms may seem incredible, there is no question that this is precisely what these structures were, and the case is not without precedents. Several years ago, for example, some fish fossils were discovered that had the microscopic detail of their body muscles preserved. Vertebrate skeletal muscle is characterized by fine striations that can be seen in the cells under high magnification, and such striations could be seen in the fossilized muscle tissue. We have referred to this as "fossilized muscle," but it is not the actual tissue that has been preserved; rather, it is a high-resolution cast of the original tissue.

How do we envisage that such fine preservation has occurred? As we said in Chapter 4, there is much that we do not understand about the process of fossilization, but we can speculate how fine details like the structure of cells can be preserved. A technique often used by botanists for studying cell structure of leaves is to paint a film of nail varnish onto the surface, let it dry, then peel it off and study it under the microscope. The varnish faithfully reproduces all of the fine detail of the superficial cells of the leaf, and, in a similar way, we may visualize that the particles which coated the muscle cells of the fish became consolidated, reproducing their microscopic structure. Such preservation, while unusual, is nevertheless comprehensible.

The age of the Gunflint stromatolites is about 2 billion years, but the oldest stromatolites date back to 3 billion years ago. Microfossils have been identified in more than forty stromatolites from different localities, most of these discovered in the last decade. The oldest fossil that has been reported, though, which has been identified as a bacterium, is dated at 3.4 million years.

The major feature that distinguishes prokaryotes (cells without a nucleus) from eukaryotes (cells with a nucleus), as we saw in Chapter 4, is size. The former are much smaller than the latter, and while there is some overlap in their size ranges, we can safely assume that cells larger than 100 micrometers are eukaryotes. Several thousand microfossils from eighteen widely separated Precambrian localities have shown that eukaryote-sized cells do not appear until about 1500 million years ago. We therefore have evidence that there was a sequential appearance of unicellular organisms, those lacking a nucleus appearing first, and thriving for about 2 billion years before the appearance of nucleated cells. There was then a period of about 800 million years before the first multicellular fossils appeared.

Why should it take such a long time for cells with a nucleus to appear? The interval is about twice as long as that between the appearance of the first cell with a nucleus and the first multicellular organisms. The answer probably lies in the fact that the most primitive cells, the prokaryotes, do not reproduce sexually. Without sexual reproduction there is no shuffling up of genes prior to cell division and therefore very little genetic variability. This in turn means that "offspring" are almost identical to their "parents"; hence the rate of evolution is slow. (Mutations do occur, of course, and these are the only source of genetic variability in asexual organisms.)

So far we have seen that

- The earliest fossils — 3-billion-year-old cells — are the simplest.
- Cells with a nucleus probably did not appear until 1.5 billion years ago.
- The first multicellular fossils are jellyfish and worms, which are among the simplest of animals. They appeared about 700 million years ago, towards the close of the Precambrian Period.

We have also seen that the Cambrian Period is characterized by the appearance of a wide diversity of marine organisms. Indeed, Dr. Gish tells us that "every one of the major invertebrate forms of life have been found in Cambrian rocks." Is he right? Before answering this question we should just note that he avoids any mention of the first appearance of the major vertebrate groups. He has good reason for doing this, of course; not even a creationist would try and tell us that amphibians and reptiles and birds and mammals have been found in Cambrian rocks.

With the exception of the bryozoans, a group of small animals which form colonies looking something like seaweed, all of the major invertebrate groups (phyla) are represented in the Cambrian. We should notice, however, that these Cambrian animals are all aquatic; land animals did not appear until the Devonian (see page 101 for a table of geological time periods). Therefore, although the major invertebrate groups are represented in the Cambrian, there are major divisions within these groups which had not yet appeared. The insects, for example, which today are the most abundant of all animals, did not appear until the Devonian Period, some 160 million years after the start of the Cambrian. There are also major divisions of marine groups that did not appear until after the Cambrian: bryozoans — Ordovician; echinoids (sea urchins) — Ordovician; crinoids (feather stars and sea lilies) — Ordovician; belemnites — Carboniferous; ammonites — Triassic.

If the major vertebrate groups are added to our geological survey, the creationists' position looks even worse. The earliest vertebrates, primitive, jawless fishes, appeared in the Ordovician (there is evidence, however, in the form of isolated fragments, that they appeared in the preceding Cambrian), and land forms did not appear until considerably later: amphibians — Devonian; reptiles — Carboniferous; mammals — Triassic; birds — Jurassic. Within each of the major divisions (classes), however, are major subdivisions (subclasses, orders, and suborders), and most of these appeared after the first appearance of their class representatives. To list these major subdivisions would be tedious, so we will only present a sampling. Amphibia: frogs — Jurassic; salamanders — Cretaceous. Reptiles: dinosaurs — Triassic; pterosaurs — Triassic; ichthyosaurs — Triassic; plesiosaurs — Jurassic; snakes — Cretaceous. Mammals: marsupials — Cretaceous; insectivores — Cretaceous; rodents — Paleocene; bats — Eocene; elephants — Oligocene. Birds: penguins — Eocene; parrots — Miocene. So far we have only listed the first appearance of animals; if we now add the plants, it reads like a floral requiem to creationism: Psilophytes (the earliest land plants) — Silurian; mosses, ferns, and horsetails — Devonian; conifers — Carboniferous; ginkgos — Permian; cycads — Triassic; flowering plants — Cretaceous.

There is no escaping the facts documented in the fossil record: the earliest organisms are the simplest, and there is a sequential appearance of the various major groups of organisms throughout time. The time period over which this unfolding of greater complexity occurs is vast, approximately 3.4 billion years.

The creationists deny that there is a sequential appearance of fossils, arguing that all organisms that are today found as fossils once lived together contemporaneously. The lengths to which they go to substantiate this claim even go beyond ignoring fossil evidence, and beyond arguing that the sequence of fossils in the rocks is a result of the Noachian flood. We will now examine two of their flights of fancy — their evidence that *Archaeopteryx*, the earliest bird, and man were contemporaneous, and that man and dinosaur once walked together.

In the introduction to this book I said that one of the reasons for using quotes from the creationists is that some of the things they say are incredible. Their evidence for *Archaeopteryx* and man living together is one of the best examples of this. Dr. Morris, quoting from an article in *Science Digest*, tells us that José Diaz-Bolio, a Mexican archaeologist-journalist, discovered an ancient Mayan sculpture of a peculiar bird with reptilian characteristics in Veracruz, Mexico. The article goes on to say that the discoverer believed that the sculpture was not merely a product of a Mayan flight of fancy but a realistic representation of an animal that lived during the period of the Mayan civilization — 1000–5000 years ago. Reference is made to *Archaeopteryx*, "to which the sculpture bears a vague resemblance . . ." What does Dr. Morris conclude from this? "The evidence seems clear that archaeopteryx, or some equivalent ancient bird, was contemporaneous with man and only became extinct a few thousand years ago." By the same token the bizarre creatures with human bodies and animal heads that are painted in Egyptian tombs must also have lived contemporaneously with man!

I first heard the story that human footprints had been discovered alongside those of dinosaurs three years ago when I read my first book on "Creation Science." The footprints in question are found along the Paluxy River in Texas, a locality famous for dinosaur tracks since Roland Bird of the American Museum of Natural History excavated there in the 1930s. I was unimpressed with the evidence presented in the book. What, after all, can be proved from a few photographs?

A year later I met a creationist who had visited the site. He said that although it was well known that the locals had, from time to time, carved human footprints in the rocks, there were other human prints that had been naturally formed. He knew these were genuine because he had been there when fresh tracks had been uncovered. A film of the operation had been made called *Footprints in Stone*, made by the Films for Christ Association, and he kindly lent me a copy.

I was unmoved by the folksy witnesses who were interviewed and I was unimpressed with most of the "human" footprints that were presented as evidence. Nevertheless some indeed did look remarkably human, even to details like having a deep, ball-like impression for the heel, and an imprint of the big toe.

Just imagine what a revolution our ideas on human evolution would have to undergo if they really were human footprints! However, I know from long experience that things do not always look the same on film as they do in reality. Furthermore, footprints can be very misleading if one is not careful. If the print is good and deep it is less likely to be misinterpreted, but so many footprints are shallow, and it is often difficult to be sure of their true shape.

A few years ago now we put a slab of Triassic dinosaur footprints on display in our gallery at the Royal Ontario Museum. The slab was crisscrossed with shallow prints made by fairly small dinosaurs, and while these could be delineated quite easily when a light was held close to the surface, or when they were dampened, it was otherwise difficult to determine their exact shape. The "human" footprints in the film had obviously been wetted, making them stand out against the rest of the rock, and I was interested to know whether there had been any bias, intended or otherwise, in the wetting process. The only solution to the problem, of course, was to fly down to Texas and examine the prints for myself, but teaching commitments prevented me from doing this.

In the meantime, I came across an article on the footprints written by biologist David Milne. He argued that the "human" footprints were actually just the single toeprints of theropod dinosaurs (theropod dinosaurs, like birds, made a three-pronged footprint, often with a dimple in front of each toeprint made by the claws), and that the other two toeprints were missing. He showed several photographs of these supposed human footprints from a paper by Dr. Morris. In one of these, sand has been used to highlight the human prints, which raises the question of whether the sand was also obscuring other details, such as imprints of side toes. I viewed the Paluxy River film again and, sure enough, in one of the "human" prints I could discern a side toe.

At this point, I went through our collection of dinosaur trackways and found a theropod imprint of about the right size. The middle toe was a bit too small to match my foot, but I found a young friend whose foot fitted perfectly. His heel fit neatly into the depression formed by the claw on the dinosaur's middle toe, and his big toe matched a depression in the dinosaur's pad just perfectly. My young friend assured me that his foot felt just right in the middle toe print, and it certainly looked like a good fit.

I conclude that the genuine "human" prints from the Paluxy River site are really just dinosaur footprints which have been misinterpreted. There is therefore no evidence that man and dinosaurs were contemporaneous.

We have seen that the fossil record provides irrefutable evidence that there was a sequential appearance of the various groups of organisms through time, and this fact alone demolishes the creationists' case. In the next two chapters we will review further fossil evidence for evolution: the occurrence of intermediate forms linking major groups.

TEN

From Reptiles to Birds: When Is a Bird Not a Bird?

 THE CREATIONISTS TELL US that there are no intermediate fossils linking major groups of organisms. *Archaeopteryx*, held by evolutionists to be a link between reptiles and birds, they say is nothing more than a bird. We will examine *Archaeopteryx* in some detail and, in comparing its anatomy with those of reptiles and birds, we will see just how wrong the creationists are.

In 1861, just two years after the publication of Darwin's *Origin of Species*, a solitary feather was found in the Solnhofen limestones of southern Germany. Aside from the fact that the limestone was Jurassic in age (about 120 million years old), making this the earliest bird fossil, this find was of no particular consequence. But then, just two months later, a second specimen was found in the same locality, this time an almost complete skeleton. The animal was obviously a bird because it had feathers and wings, but a close examination revealed a number of reptilian features, too. The skeleton was named *Archaeopteryx*, meaning ancient wing.

It had long been recognized that birds and reptiles were closely related because they share a number of features, mostly internal, but including the possession of scales and the laying of a shelled egg, and here was tangible evidence of the connection. Thomas Henry Huxley, the evolutionist who crossed swords with Bishop Wilberforce in the great evolution-creation debate at Oxford in 1860, recognized that *Archaeopteryx*, with its mixture of reptilian and avian features, was an important piece of evidence in the evolutionists' cause. His anti-evolutionary contemporaries, however, failed to see the connection, and the German paleontologist J. A. Wagner dismissed *Archaeopteryx* as a feathered reptile. With total disregard for the biological law of priority, which states that once an organism has been given a proper name it cannot be subsequently renamed, he renamed it *Griphosaurus problematicus* — the puzzling reptile.

Huxley set himself the task of determining which particular reptilian group had most in common with birds, and concluded that it was the dinosaurs. At that time there were far fewer dinosaurs available for study than there are today, and the fact that almost all paleontologists now agree that birds did evolve from dinosaurs is a fitting tribute to Huxley's acumen. There are, in fact, two basic types of dinosaurs, called ornithischians and saurischians, each with various subdivisions. Birds are most similar to the theropods, which belong to the saurischian group. Most theropods were carnivores and the group includes *Tyrannosaurus* and *Allosaurus*. All discussions that follow pertain to theropod dinosaurs.

Anyone who has been fortunate enough to examine closely a dinosaur skeleton could not fail to be impressed by its startling resemblance to that of a bird. Both, for example, have a three-toed foot, each toe ending in a sharp claw, usually with a backwardly projecting big toe (the spur, in a bird). Just below the ankle regions are three rod-shaped bones, the metatarsals, which are tightly pressed together in dinosaurs and fused in birds. Even the specialized arrangement of bones in the ankle joint is the same in birds and theropods and is unique to these two groups. Because the small bones of the ankle joint are fused with the larger bones in adult birds, we have to look at immature birds to be able to see them.

Moving up from the ankle we come to the shinbone, the tibia, and its partner, the fibula. Here again the resemblances between birds and dinosaurs are impressive, right down to the details of the crests on the tibia. The major difference between the two is that the fibula is reduced in modern birds, and does not extend all the way to the ankle joint. The thigh bone (femur) of dinosaurs is closely similar to that of birds, and a particularly interesting feature is that the head — the knob that fits into the hip socket — is set off from the shaft of the bone. This condition, which is not seen in any other reptiles,

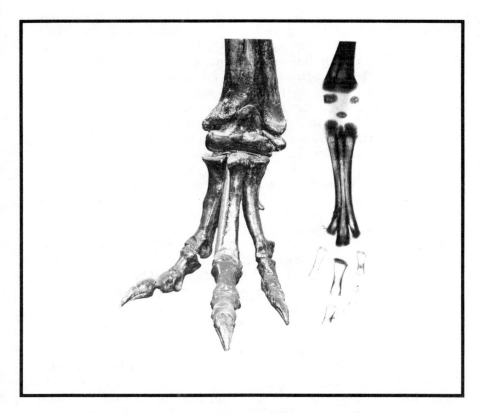

Foot and lower leg of a theropod dinosaur (left) and an immature bird.

allows the leg to be placed vertically beneath the body, giving the animal an erect posture. Other reptiles have their legs splayed out at the sides as if they are doing push-ups. Birds and dinosaurs have similar hips (pelvis), though in most birds, the more advanced ones, that is, the overall similarity is partially obscured because of fusion between the individual bones. However, the structure shows up clearly during development, as in the case of the ankle region.

The similarities between dinosaurs and birds are so obvious that some paleontologists have even suggested that they should be classified in the same major grouping. Most of us think that this is a bit extreme, though, because there are significant differences between the two, which we will now look at.

The pelvis of theropod dinosaurs is a three-pronged structure when viewed from the side. The front prong, which usually slopes forward,is formed by the pubis and, towards its tip, it

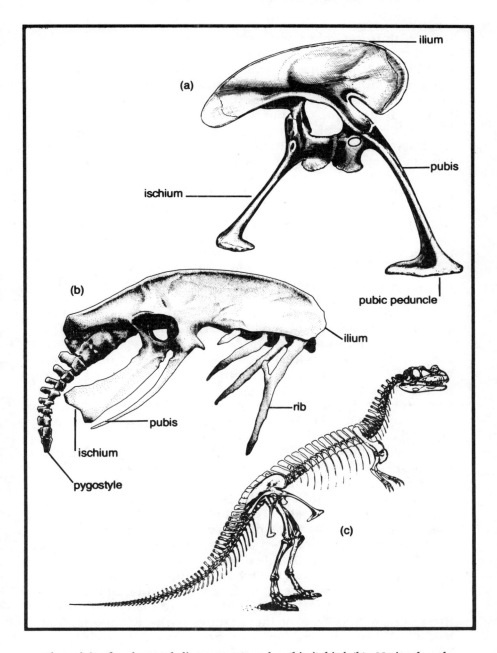

The pelvis of a theropod dinosaur (a) and a (kiwi) bird (b). Notice that the pubis points backwards in the bird and that it lacks a pubic peduncle. The skeleton of a theropod dinosaur (c). Notice the long, bony tail, which contrasts with the short tail of the kiwi.

fuses with its neighbor of the other side to form a large knob, called the pubic peduncle. In birds the pubis, which projects backwards, lacks a pubic peduncle. This major difference in pelvic structure is probably correlated with weight distribution and balance. Dinosaurs counterbalanced the weight of their body with their long tail, but birds, having almost no tail at all, have shifted their heavy internal organs well back, to lie on a level with the hip joint. Although birds have a small tail skeleton, many of them, like the magpie, pheasant, and peacock, have very long tails. The tail, of course, is formed only of feathers and these are attached at their base to a fleshy structure, supported internally by the tail skeleton. You see this structure every time you prepare a turkey for the oven — it is that fat flap that you lift up to stuff the bird — sometimes called "the parson's nose." Have a good look at it next time; if you cut it open, preferably after it has been cooked, you will see that it contains a great deal of fat. This is where the bird gets the oil to preen its feathers. If you cleaned all the meat and skin from the tail of your cooked turkey, you would see that the tail skeleton ended in a flattened wedge of bone, called the pygostyle. The pygostyle is found only in birds.

Now that we have our hands all greasy we might as well take a look at another unique feature of birds, and for this we need to remove one of the bones from the neck, or from the back region. The individual bones, or vertebrae, fit together in such a way that there is a limited amount of movement between them, and the actual surfaces of contact are quite unlike those of any other animal. If we could look at our own vertebrae we would see that the contact, or articular, surfaces were quite flat, but in birds the surfaces have a complex double curvature, very much like that of a saddle. No other animal has saddle-shaped articular surfaces between its vertebrae.

When you carve off those big slices of white meat from the breast of the turkey you are cutting through the main flight muscles, and these are attached to the breastbone or sternum. All birds have a sternum, though it is reduced in the flightless birds, but dinosaurs have no sternum at all (though they may have had a sternum of cartilage like that of many living reptiles). Just at the front of the sternum is the wishbone, that V-shaped bone we used to fight over when we were children. The wishbone, or furcula, which is probably equivalent to fused collarbones (clavicles), is found only in birds.

So far we have seen that birds differ from dinosaurs in their hips, tail, vertebrae, and chest region, but there are several other differences that we still have to consider. Dinosaurs typically have three fingers in the hand, corresponding to our thumb, index, and middle fingers (designated fingers 1, 2, and 3), with two, three, and four bones in each respectively. Each finger ends in a prominent claw. These three fingers articulate with three separate and well-developed bones, called metacarpals (1, 2, and 3), which correspond to the palm of our hand. Birds have one main finger — the second — the first finger being reduced to a small structure called the alula, or bastard wing, while all that remains of the third finger is a small nubbin of bone. There is also much modification in the metacarpal bones; the first is barely visible, and the second and third are fused into a distinctive structure (called a carpometacarpus) found only in birds. Some birds have small claws at the ends of one or two of their fingers, but most do not (see page 124).

The avian ankle, as we saw earlier, is closely similar to that of dinosaurs, even to the details of the small bones, but there are two differences in the bones below the ankle. First, the three metatarsal bones, as we have seen, maintain their separate identities in dinosaurs, but in adult birds they are completely fused to form a single bone in which they can no longer be distinguished. Secondly, at the back of this bone is a protuberance called the hypotarsus, which is often quite well developed and complex in shape. There is no hypotarsus in dinosaurs.

The last feature we have to consider is the abdominal ribs. Theropod dinosaurs, like many other reptiles, have a second set of ribs in the abdominal region, but birds do not have abdominal ribs. The following table lists the differences between reptiles, as exemplified by theropod dinosaurs, and modern birds.

We could also add the point that dinosaurs have teeth while birds do not, but there are some birds, of Cretaceous age (about seventy million years ago), which do have teeth; therefore, this is not such a useful feature in distinguishing reptiles from birds. While this might seem a formidable list of differences, we should remember that there are still so many similarities between the two groups that some specialists advocate uniting them within the same class.

Aside from the solitary feather found in 1861 there are five specimens of *Archaeopteryx*. Three of these are complete or near-complete skeletons, located in London, East Berlin, and Munich, and although there are a few points that remain unresolved, we have a sound knowledge of the anatomy. I have

	Reptilian condition	Avian condition
HIP	1. Pubic peduncle present.	No pubic peduncle.
TAIL	2. Long, bony tail.	Short, bony tail.
	3. No pygostyle.	Pygostyle present.
VERTEBRAE	4. Articular surfaces not saddle-shaped.	Articular surfaces saddle-shaped.
CHEST	5. No bony sternum.	Bony sternum present.
	6. No wishbone.	Wishbone present.
HAND	7. Usually three well-developed fingers.	Only one well-developed finger (2nd), thumb reduced to small size, finger 3 not visible.
	8. Three well-developed metacarpal bones.	Two well-developed metacarpal bones.
	9. Metacarpal bones unfused.	Metacarpal bones fused.
ANKLE REGION	10. Metatarsal bones separate.	Metatarsal bones lose separate identity.
	11. No hypotarsus.	Hypotarsus present.
ABDOMEN	12. Abdominal ribs present.	No abdominal ribs.
FEATHERS	13. Feathers apparently absent.	Feathers present.

had the opportunity of examining the London and East Berlin specimens in detail, but there are so many excellent accounts of the anatomy of *Archaeopteryx* that a visit to Europe is not a prerequisite for an enlightened discussion.

In the next table we work our way down our list of characteristics, noting whether they are in the reptilian or the avian condition in *Archaeopteryx*.

In the majority of its features *Archaeopteryx* is clearly reptilian, but what do the creationists tell us about it? Dr. Morris tells us the "*Archaeopteryx* is a bird, not a reptile-bird transition," and this certainty is echoed by Dr. Gish: "It was not a half-way bird, it *was* a bird." Creationists, as we have seen before, do not hold back from making authoritative pronouncements upon subjects beyond their expertise.

The only avian features of *Archaeopteryx* are the possession of feathers and of a wishbone. Do these two features represent such a big jump from the dinosaurian condition? Certainly not in the case of the wishbone, which, after all, just involves the fusion of the two collarbones. Feathers, however, are another matter, and must be considered as a major evolutionary jump. The possession of feathers not only permits flight, but also

Checklist of Reptilian and Avian Features of *Archaeopteryx*

		Reptilian condition	*Avian condition*
HIP		1. Pubic peduncle present.	
TAIL		2. Long, bony tail.	
		3. No pygostyle.	
VERTEBRAE		4. Articular surfaces do not appear to be saddle-shaped (caution is required here because only two articular surfaces can be seen).	
CHEST		5. No bony sternum.	
		6.	Wishbone present.
HAND		7. Three well-developed fingers (each with the same number of bones as in most dinosaurs).	
		8. Three well-developed metacarpal bones.	
		9. Metacarpal bones unfused.	
ANKLE		10. Metatarsal bones separate.	
REGION		11. No hypotarsus.	
ABDOMEN		12. Abdominal ribs present.	
FEATHERS		13.	Feathers present.

insulates the body, making it possible for high body temperatures to be maintained. Many evolutionists consider that feathers were evolved — by the modification of scales — primarily for their insulating properties, and only later became aerodynamic structures. The question is, of course, how something as complex as a feather could possibly have evolved from something as simple as a reptilian scale, and this raises that old problem of how an intermediate structure could be of any use to an organism. I believe that both problems are surmountable.

Most familiar reptiles, like the lizards and snakes that some of us used to keep when we were young, have scales that lie flat against the surface, but there are others, like the spiny-tailed lizard, where the free border of each scale is drawn out into a prominent projection which lies above the surface of the skin. Dr. P. J. Regal has shown that elongated body scales like these have important insulative properties which are of direct benefit to the animal. He goes on to argue that scales which are subdivided (as when a series of parallel cuts are made down both sides of a sheet of paper), being flexible, would have better insulating properties than individual scales. It is not diffi-

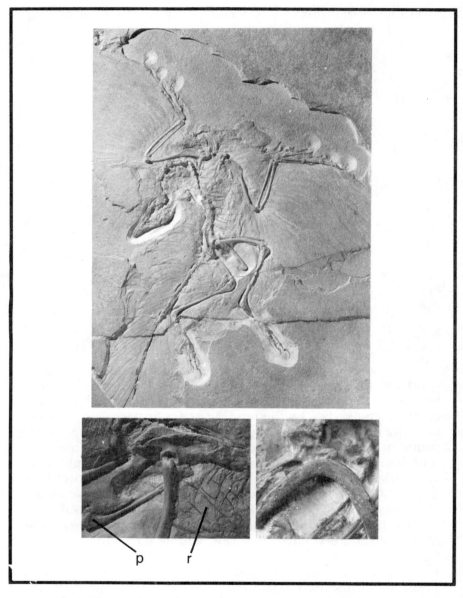

TOP: *The Berlin specimen of* Archaeopteryx. *Notice the long, bony tail and the well-developed clawed fingers.*

BOTTOM LEFT: *Detail of the pelvis. Notice the pubic peduncle (p) and the abdominal ribs (r).*

BOTTOM RIGHT: *The wishbone of the London* Archaeopteryx *specimen.*

cult to imagine how individual variation, directed by natural selection, could lead to the evolution of increasingly more complex scales, scales that begin to look more and more like primitive feathers.

Drs. Morris and Gish both ask why we do not find fossils with partly developed feathers, and I believe that the answer lies in the incomplete nature of the fossil record. The fact that feather impressions can be clearly seen in two of the five skeletons of *Archaeopteryx* might lead one to conclude that the preservation of such fine detail is not exceptional, but this is most certainly not true. First let us point out that we only have knowledge of *some* of the feathers of *Archaeopteryx*, namely the largest ones: the primary (on the hand) and secondary (on the forearm) wing feathers, and the tail feathers. We have no knowledge at all of the numerous smaller feathers, called contour feathers, which fill in gaps between the large wing feathers, and which clothe the body. Secondly, the feather impressions are so faint in two of the specimens of *Archaeopteryx* that they were overlooked and the specimens were initially identified as reptiles. Let us be in no doubt — feather impressions *are* rare in the fossil record, and even when they have been preserved they only give us information about some of the feathers. The only place left to search for evidence of the transition from scales to feathers is in the living world.

If we examine the wing of a penguin, we see a wide range of covering structures, from small structures that look like scales at the leading edge, to structures that are obviously feathers at the trailing edge. There are all shades in between. Before making a closer inspection we need to be familiar with the anatomy of a regular feather.

A feather has a central rib, called the rachis, which bears numerous side branches, called barbs. The barbs, in turn, bear small side branches, called barbules, and these are provided with hooks which interlock with those on the adjacent barbules. In this way the barbules are held together in a continuous web, called the vane. When a feather is ruffled, by stroking it the wrong way, the interlocking barbules are separated, but the feather can be restored by stroking it the other way, which causes the barbules to interlock once more.

Viewed under a microscope the scale-like structures at the leading edge of the penguin wing can be seen to be fringed by long processes that project from, and are continuous with, the margin. The structure could be simulated by making a series of diagonal cuts along the margin of an oval piece of card. Closer

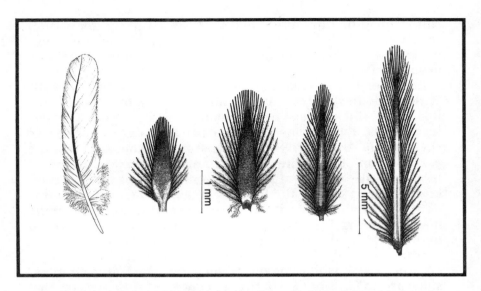

Comparison of the wing feathers from a flying bird (far left) with the wing feathers of a penguin. The first penguin feather was taken from the leading edge of the wing, the last penguin feather from the trailing edge. Notice that the feathers become less scale-like towards the trailing edge.

examination of these long processes reveals that they have smaller processes projecting from their margins. These fine processes are not very obvious, because they lie flat along the margins of the larger processes. There are obvious resemblances between this scale-like structure and a regular feather: the long processes correspond to the barbs, and the fine processes to the barbules, while the central scale corresponds to the rachis. The proportions, however, are markedly different; the barbs comprise the largest portion of the area of a regular feather, but the rachis predominates in the scale-like structure.

Whether the scale-like structures are interpreted as modified scales or as modified feathers (that is, whether they are considered primitive or highly specialized) is unimportant to us here. What is important is that they illustrate that feathers and scales are essentially just variations on a theme; both are formed of a horny protein called keratin, and they both develop along similar embryonic pathways. They also illustrate how intermediate structures, in this case half-feather half-scale, can be fully functional.

I must emphasize that I am not suggesting that penguins are intermediate between reptiles and birds, merely that they have the ability to produce a wide spectrum of covering structures, from scales (on their legs), through feather-like scales, all the way to regular feathers. I consider this to be further evidence that birds evolved from reptiles, and we will now look at another piece of evidence, this time from embryology.

We mentioned earlier that *Archaeopteryx* has teeth, but we did not include this feature in our list simply because certain Cretaceous birds also have teeth. Dr. Gish draws attention to the same point, referring to this "alleged reptile-like feature" of *Archaeopteryx* with disdain because it is found in later birds. He misses the important point, though, that the possession of teeth is a primitive feature which *Archaeopteryx* and Cretaceous birds inherited from their reptilian ancestors. Modern birds do not have teeth, but a recent experiment on developing embryos has shown that the chicken, at least, has retained the *ability* to form teeth. In this experiment, small pieces of epithelial (covering) tissue were taken from 5-day-old chicken embryos and combined with tissue taken from the tooth region of 16- to 18-day-old mouse embryos. The combined patches of living cells were incubated for several weeks, then examined under a microscope. It was found that partial or complete teeth were formed. The fact that no teeth could be formed without adding the chick cells to the mouse cells shows that tooth formation was actually initiated by the bird cells. This in turn shows that the cells of the chicken retain the ability to direct the formation of teeth — that is, they retain the genetic information for tooth formation — but teeth are not normally formed because other factors are missing. Perhaps all birds have retained the structural genes necessary to form teeth, but these may be kept in the "switched-off" condition by certain regulatory genes.

Gene regulation, as we mentioned in Chapter 3, is of considerable evolutionary importance because it permits genes to be passed on from one generation to the next without necessarily being expressed. Small changes in regulator genes could then bring about rapid evolutionary changes by causing latent structural genes to be turned on. Regulator genes could also have profound effects on the embryonic development of organisms by changing the timing of certain events. For example, it has long been recognized that our own adult bodies have more in common with immature primates than with adult primates, both being hairless and having relatively large heads, to mention just two similarities. This led many biologists to speculate that we may have evolved from the same hairy primate ancestor through the suppression of the last stages of development, and by the acceleration of the onset of sexual maturity. This process is called neoteny. Support for the importance of gene regulation during development has come from recent studies of the molecular structure of the proteins (the assorted and fairly complex molecules from which bodies are built) of man and great apes. It has been found that the proteins are strikingly similar, more similar even than the proteins of sibling species (see Chapter 2). This implies that apes and man have very similar structural genes, but, as a visit to the zoo would confirm, there is a world of difference between our anatomy and that of an ape. These major differences can be explained in terms of the differences in regulatory genes. To give a hypothetical example, if during the development of a gorilla the structural genes controlling the formation of brow ridges were turned off, the resulting individual would not have the beetling brows that characterize his kind.

Having discussed the idea of gene regulation, we can understand the relationship between embryos and ancestors; this will allow us to assess some additional evidence for the close relationship between reptiles and birds. Most organisms start off life as a fertilized ovum, and from this they eventually develop into a fully formed adult. The first phase of development, called embryology, usually takes place inside an egg, or, in the case of most mammals, inside the uterus. This is followed by a stage of growth, and eventual maturity, which follows after hatching or birth. Considerable attention has been given to the study of embryos, especially by evolutionists, who have used this evidence to support the theory of evolution. Some early embryologists, working during the last century, were impressed by the way that an embryo starts off like a single-celled animal. From this it becomes a cluster of cells, then a hollow ball, and these various stages that it goes through in some ways resemble a series of (adult) multicellular organisms arranged in order of increasing complexity. And so was born the old idea of recapitulation — that embryos go through all the major stages of their evolutionary history during their development. Like so many ideas, it seemed like a good one at the time, but, as the creationists like to point out to us, the idea has long since been rejected. Just because the theory of recapitulation has been rejected, though, there is no need to reject embryonic evidence out of hand. We will now see that embryology provides us with some persuasive evidence that birds evolved from reptiles.

When we were looking at the similarities between dinosaurs and birds we saw that both groups had very similar ankle regions, but that this similarity is obscured in adult birds by the fusion of the individual bones. Take, for example, the three metatarsal bones. In most theropod dinosaurs these remain distinct throughout their length, even if they do become partially fused, and the same is true for *Archaeopteryx*, but in adult birds it is impossible to see the joints between them because they become fused together into a single bone. If birds were created independently, why should this structure not be a single bone throughout all stages of its development, instead of being formed from the fusion of three separate elements? As an evolutionist I interpret the evidence as follows. Adult individuals of theropods and of *Archaeopteryx* have three metatarsal bones, and if it were possible to look at their embryonic development I would expect to see that these three bones developed from three separate elements. Modern birds inherited the capacity to form three separate metatarsals during their embryonic development, but these elements become joined together during the later stages of development. The same reasoning is applicable to the small bones in the ankle joint, which remain separate in adult theropods and in embryonic birds, but which become fused with the leg bones in mature birds.

If we now look at the early embryonic development of the wing of a bird we see that there are three separate fingers, each supported by a separate metacarpal bone (metacarpals 1, 2, and 3). The capacity to form three separate fingers has been inherited from their reptilian ancestors. As development proceeds, metacarpal 3 becomes markedly bowed, and eventually fuses with metacarpal 2. The first digit (the alula) remains separate, though it is so small in most birds that it can be seen only on the closest inspection. The evidence that birds are the modified descendants of theropods is compelling.

One of the most obvious features of *Archaeopteryx*, setting it apart from modern birds, is the presence of three clawed fingers on the wings. Evolutionists interpret the fingers of *Archaeopteryx* as evidence that it evolved from reptiles, but Dr. Gish rejects this on the grounds that there is a bird living in South America, the Hoatzin, in which juveniles have claws on the wings. He goes on to argue that the Hoatzin is "100% bird" but that it possesses a character "used to impute a reptilian ancestry to *Archaeopteryx*." In saying this he shows us that he has entirely missed the point that the young Hoatzin, in retaining a primitive reptilian feature which other birds lose just before leaving

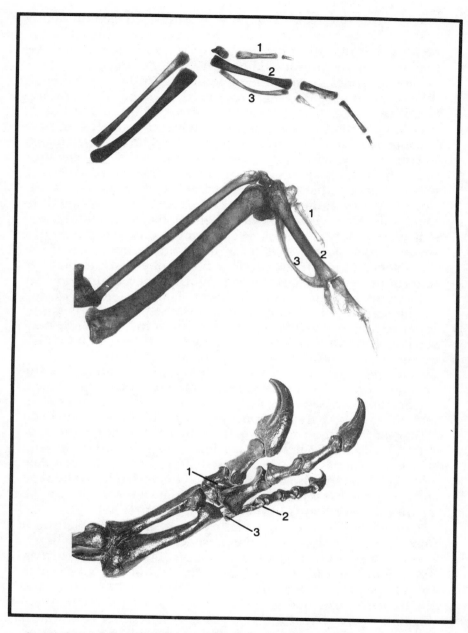

Comparison of the forelimb of an immature bird (top) with that of an adult bird (middle), and of a theropod dinosaur (bottom). Notice that all three have three fingers and three metacarpals (1, 2, 3), but that metacarpals 2 and 3 have become fused in adult birds. (The bird featured is actually a Hoatzin.)

the egg, is showing us its reptilian pedigree. Far from being evidence to the contrary, the Hoatzin is additional evidence for the reptilian ancestry of birds.

Before leaving the wing of *Archaeopteryx* we should say something more about one of the standard creationist arguments about the value of half-formed structures. (We have already seen that half-scale half-feather structures are fully functional in the wing of a penguin.) What is the value, asks Dr. Morris, of a half-formed wing, or of any other half-formed structure? He concludes that incipient features have no selective advantage at all, and might even be harmful until fully developed. We all recognize that the forelimb of *Archaeopteryx* is a wing, but what in fact is the difference between it and the forelimb of a dinosaur? The only significant difference is that *Archaeopteryx* had long feathers, whereas dinosaurs apparently did not. It is more than likely that feathers evolved as heat insulators, only later evolving into aerodynamic structures (such features, which are originally evolved under one particular selection pressure only to evolve further under an entirely different one, are usually called preadaptations, but the new term *exaptation* is preferred.) Therefore, it is difficult to visualize how a feathered arm could possibly have been disadvantageous. Furthermore, when we consider the obvious advantage that gliding animals such as flying squirrels, flying lizards, and flying frogs derive from webs of skin that act as parachutes, it is obvious that a gliding animal having progressively more extensively feathered forelimbs would have an increasing advantage over its fellows.

Dr. Gish tells us that *Archaeopteryx* was a *true bird* because it had wings, it was completely feathered, and it flew, but his assertions are questionable. First, we do not know that *Archaeopteryx* flew, nor do we know that it was completely feathered, though both suggestions seem reasonable. Secondly, we have to remind him that there are plenty of true birds living today that are completely feathered and have wings but do not fly (ostriches, cassowaries, emus, rheas, kiwis, flightless cormorants, steamer ducks, and various rail species, to mention just a few). Furthermore, until recently there was a group of birds living in New Zealand, the moas, which had feathers but no wings. Nobody would suggest that non-flying birds, or the wingless moas, were not birds. The obvious feature these birds all share is the possession of feathers. If birds are defined as animals with feathers, as Dr. Gish seems to imply, then it is obvious that intermediates have no reality. An animal either has feathers, and hence is defined as a bird, or is without feathers, in which case it is defined as a non-bird. The penguin is an interesting case because it has feathers and half-feathers and, to make things more complicated, it does not fly. Should it be called a true bird, a non-bird, or perhaps an ex-bird? This conjures up images of a famous Monty Python sketch about an ex-parrot — perhaps it is time to move on to the next chapter, which is concerned with a second example of intermediate fossils: the fossils that bridge the gap between reptiles and mammals.

ELEVEN

A Mammal by Any Other Name

CREATIONISTS TELL US THAT there are no intermediate fossils linking major groups of organisms. In the previous chapter we examined the link between reptiles and birds; in this chapter we will look at fossils that link reptiles with mammals.

We saw in Chapter 6 that many of the significant changes that occurred in the evolution of mammals relate to features that are not preserved in the fossil record, and many of these features are used to identify modern mammals. They include: the maintenance of relatively high and constant body temperatures, the possession of fur, mammary glands, and sweat glands, the placental development of the young inside the mother, and extended parental care. Drs. Gish and Morris both acknowledge this fact, and the latter goes so far as to say that reptiles and mammals are so similar in their skeletal anatomy that "their fossilized remains provide little basis for distinguishing between them." This is not true, because there are many ways of distinguishing between the skeletons of fossil reptiles and mammals. The difficulty comes, as we have already seen, when we try to distinguish between transitional

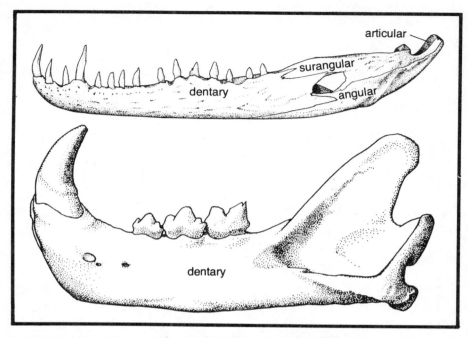

Comparison of the lower jaw of a crocodile (top) with that of a jaguar. Notice that the crocodile teeth are essentially all the same and lack cusps, whereas the jaguar teeth vary in shape and have complex cusps. The crocodile has several bones in its jaw, whereas the jaguar has only one bone, the dentary.

forms. These transitional fossils, which link major groups, are of considerable interest to us here because they provide compelling evidence that one group evolved from the other.

Before looking at some of the fossils that are intermediate between reptiles and mammals, we should take a look at the major skeletal differences between these two groups. For our examples we will choose a crocodile and a jaguar. Most of the features we will be considering concern the anatomy of the skull, and the reason for this is simply that cranial material for mammal-like reptiles is more frequently found, hence better known, than the rest of the skeleton (postcranial material).

One of the most striking features about a crocodile's skull is its formidable array of sharp teeth. Each tooth is a simple conical peg, and although there is some variation in size and in relative robustness, the teeth are essentially all the same. Contrast this with the jaguar skull, where the teeth are specialized for different functions, and where the structure of each cheek tooth is complicated by the possession of cusps. At the front

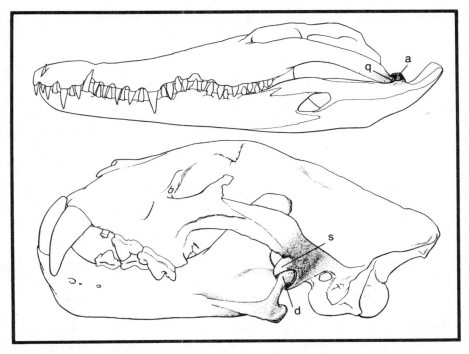

In the crocodile (top), the jaw joint is formed between the quadrate and the articular bone, whereas in the jaguar it is formed between the squamosal and the dentary bone.

end of the upper and lower jaws are the chisel-shaped incisor teeth, used for nipping and tearing. Moving back from the tip of the snout, the next teeth we encounter in the mammalian skull are the canines. These are the long stabbing teeth that the jaguar uses for killing its prey. Mammals never have more than four canine teeth, one in each half of the jaw, upper and lower. Behind the canines are the cheek teeth, and these have the most complicated structure of all the teeth owing to the various cusps. They are used by carnivores, like the jaguar, for cutting and slicing, though many other mammals, man included, use them for grinding and chewing. We can actually distinguish two types of cheek teeth, premolars and molars. The premolars tend to be simpler than the molars, but the most significant difference between them is that the premolars are present as milk teeth, whereas molars are not. We therefore get only one set of molars, but two sets of premolars. Reptiles, in contrast, replace their teeth throughout life, which always seems grossly unfair to me with dental fees as high as they are.

The cusps on the cheek teeth of mammals can be fairly complex, and they are so distinctive from one species to another that they are used for species identification.

If we look at the lower jaw of the crocodile we can see that it comprises a number of bones. Viewed from the outside the most obvious of these are the dentary, the surangular, and the angular. The articular, a fairly inconspicuous bone, is found at the top back corner of the lower jaw, and forms the lower half of the jaw joint. This bone is concave, from side to side, and articulates with the convex surface of the quadrate bone at the back of the skull. The jaw joint is therefore formed between the articular bone and the quadrate bone.

In contrast to the reptilian condition, the mammalian lower jaw has but a single bone, the dentary, and the part that forms the lower half of the jaw joint, instead of being concave from side to side as it is in the crocodile, is formed into a scroll-shaped process. This fits into a transverse groove (the glenoid) in the squamosal bone at the back of the skull. The jaw joint in the mammal is therefore formed between the dentary and the squamosal. Yet another difference between the two lower jaws is that in the crocodile the upper margin is essentially straight, whereas in the jaguar it is drawn up into a deep flange, just in front of the jaw joint, called the coronoid process.

Crocodiles are not great thinkers, nor are any other reptiles, which is not surprising when we consider what tiny brains they have relative to their size. The cranium (brain box) of reptiles is small, and comprises but a small part of the entire skull. When we look down at the top of the crocodile's skull we have some difficulty in locating the cranium, but there is no problem with the jaguar skull because the cranium is the most prominent feature.

The crocodilian skull is attached to the neck through a joint formed between a ball or condyle at the base of the skull and a socket at the front of the first neck vertebra (called the atlas — named after Atlas, the god who supported the world on his shoulders). This condyle is single in reptiles and is placed quite low down. If we now look at the jaguar skull, we see that there is a pair of condyles, and these are placed relatively higher up on either side of the hole at the back of the skull (called the foramen magnum, which is Latin for big hole) through which the spinal cord passes on its way to the brain. What is the significance of these differences? It seems that it all has to do with head mobility. Mammals can move their heads much more than reptiles. When we nod our head the two condyles at the

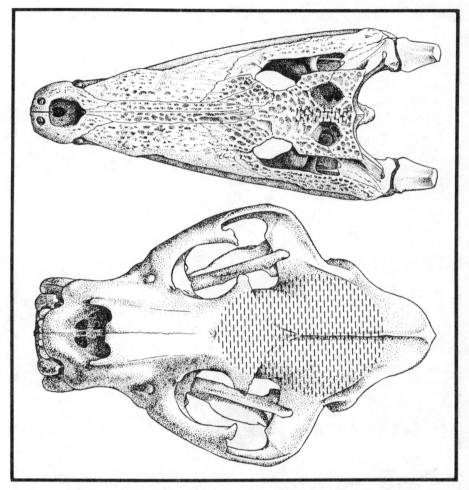

Comparison of cranium size (shaded area) in a crocodile (top) and a jaguar.

back of our skull move in the paired sockets in the atlas vertebra. If the condyles were placed low down, as they are in reptiles, the amount of flexion that our spinal cord and the hind portion of our brain would suffer would be much greater. Since our brain is relatively so much larger than the brain of a reptile, this would be undesirable. Having paired condyles is therefore really just a consequence of having a joint which is placed high relative to the opening at the back of the skull. Sideways movement of the head, as when we shake it, is brought about by the movement of the atlas vertebra on the

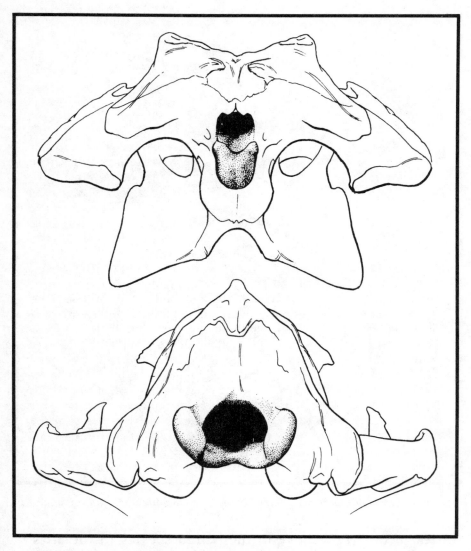

Comparison of the skull attachment in a crocodile (top) and a jaguar. Notice that the condyle (stippled area) is single in the crocodile, double in the jaguar.

second neck vertebra (called the axis). The joint between these two vertebrae is a peg-and-socket arrangement, the atlas essentially hanging on the peg (called the odontoid process) of the axis like a hat on a coathook. For those interested in the gruesome details of hanging, the sudden and unnatural up-and-down movement of the atlas-axis joint when the trapdoor is opened drives the odontoid process into the spinal cord, causing death. Reptiles, in contrast to mammals, do not have an odontoid process, and the joint between their atlas and axis vertebrae is like that between any other two vertebrae in the vertebral column. Hanging would presumably be an unsatisfactory way of dispatching a crocodile.

There are several other differences between the vertebral column of reptiles and that of mammals. Crocodiles, like other reptiles, have ribs attaching to their neck vertebrae, but these are absent in mammals (actually they are very small and have become fused to the vertebrae). Furthermore, reptiles have ribs attaching to all of the vertebrae between the shoulders and the hips, but in mammals the ribs are restricted to the chest region (the region without ribs, just in front of the pelvis, is called the lumbar region).

We mentioned in the previous chapter that most reptiles splay their legs out to the sides of their body, push-up fashion, and this can easily be seen if we look at a crocodile. Mammals, like dinosaurs and birds, place their legs vertically beneath the body. We can tell whether an animal walked with its legs splayed out at the sides or held vertically beneath the body by examining the position of the joint surfaces — joint surfaces are recognized by their smoothness. Reptiles also differ from mammals in having more finger and toe bones (phalanges) in their hands and feet. If you take a look at your own hands, you will see that there are two bones in the first digit (the thumb) and three in each of the others; the same is true of your feet. Our phalangeal count is therefore written as 2,3,3,3,3, and this is the typical mammalian pattern. Reptiles, in contrast, typically have more bones, and the phalangeal count for the crocodile hand is 2,3,4,4,3.

The last skeletal difference between reptiles and mammals that we need to consider concerns the hip or pelvis. The upper pelvic bone, the ilium, is directed upwards in most reptiles, sometimes also backwards, as in the crocodile. In the mammals, however, the ilium slopes well forwards. This is not an exclusive mammalian feature, because it does also occur in some dinosaurs and in pterosaurs, but, taken with the other skeletal features, it is useful for distinguishing between primitive mammal-like reptiles and mammals.

As we have now considered a number of skeletal features that distinguish between reptiles and mammals, it would be useful to list them.

	Reptilian condition	Mammalian condition
SKULL	1. Teeth essentially all the same.	Teeth specialized for different functions.
	2. Teeth relatively simple, without complex cusps.	Cheek teeth with complex cusps.
	3. Lower jaw comprises several bones.	Lower jaw comprises a single bone (the dentary).
	4. Jaw joint formed between articular bone and quadrate bone.	Jaw joint formed between dentary bone and squamosal bone.
	5. Jaw joint usually formed between a hollow in the lower jaw and a rounded knob in the skull.	Jaw joint formed between a scroll-like knob on the lower jaw and a hollow in the skull (the glenoid).
	6. Lower jaw usually not very deep posteriorly.	Lower jaw very deep posteriorly owing to an ascending flange called the coronoid process.
	7. Small cranium.	Large cranium.
	8. Single condyle at the back of the skull for articulation with the neck.	Double condyle at the back of the skull for articulation with the neck.
POST-CRANIAL SKELETON	9. Second neck vertebra (axis) without a peg (odontoid process).	Second neck vertebra (axis) has a peg (odontoid process).
	10. Ribs in neck region.	No ribs in neck region.
	11. Ribs prominent all along the backbone from shoulder to pelvis.	Ribs confined to chest region.
	12. Legs usually splayed from the sides of the body.	Legs placed vertically beneath the body.
	13. Number of bones in the fingers and toes usually exceeding 2,3,3,3,3.	Number of bones in the fingers and toes not exceeding 2,3,3,3,3.
	14. Upper pelvic bone (ilium) does not usually slope forwards.	Upper pelvic bone (ilium) slopes forwards.

This list, which includes just some of the skeletal features that can be used to distinguish between reptiles and mammals, clearly shows that Dr. Morris is wrong when he tells us that their fossil remains can barely be distinguished. We seldom have any difficulty in assigning a given skeleton to the reptiles or to the mammals, and when such difficulties do arise it is because we are close to the boundary between the two groups.

The reptilian group from which the mammals evolved is collectively called mammal-like reptiles, or synapsids. They appeared towards the end of the Carboniferous Period, and are therefore one of the oldest reptilian groups, but they were at their peak in the succeeding Permian, and also for much of the Triassic Period, which saw the beginning of the dinosaurs. The mammal-like reptiles were a highly successful group, and they came in all sorts of shapes and sizes. Some were plant-eaters, others fed on meat, and they ranged from cat-sized to being larger than cows. Counted in their number were the remarkable sail-fins — *Dimetrodon* and *Edaphosaurus* — animals with enormous sails on their backs, which were supported by the greatly elongated neural spines of their vertebrae. Declining in both numbers and diversity during the Triassic Period, probably because of competition from the dinosaurs, the mammal-like reptiles finally became extinct during the Jurassic, but not before giving rise to the mammals, some time during the late Triassic.

Dr. Gish, however, finds this all too much. He argues that if mammals did evolve from reptiles at such a relatively late period in reptilian history, then it would be reasonable to assume that the reptilian group that gave rise to them also developed late, whereas the reverse is true. I must confess that I cannot follow his logic. His argument that mammal-like reptiles could not have given rise to mammals because they were such an ancient group themselves is something like saying that the young Canadians that live near me could not have been born of Greek parents because Greeks have a history stretching back to ancient times.

The mammal-like reptiles are so named because they possess some features that are found in mammals, but we must emphasize that most of them are not very reminiscent of mammals at all and only one relatively small subdivision, the cynodonts, which share many features in common with mammals, are believed to have been the group that gave rise to the mammals. Some cynodonts, like *Cynognathus*, were large, about two meters long, but many were dog-sized, or smaller. I imagine

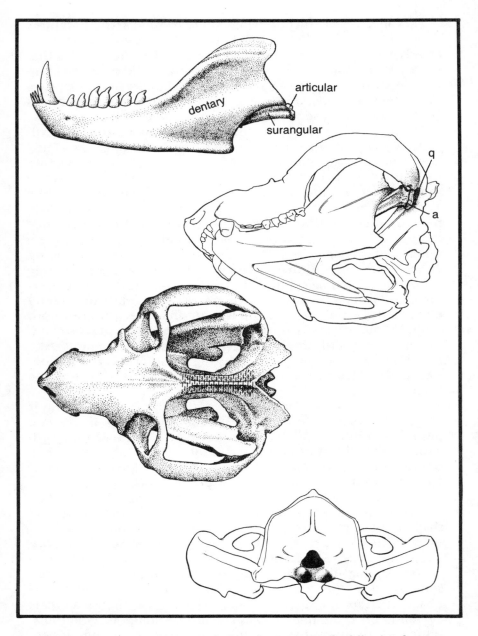

Skull features in a cynodont (**Probelesodon**). *Notice the following features: teeth vary in shape; cheek teeth have simple cusps; lower jaw comprises several bones; jaw joint is formed between the quadrate and the articular bone; cranium is small; the condyles for skull attachment are double.*

that if we could see them alive today we would think that they were mammals; they were probably hairy, but were perhaps more sluggish and less agile than most modern mammals. The cynodonts have been known for more than a century now, but it is only in recent years that enough good material has been collected to give us a fairly detailed knowledge of their anatomy. Most early accounts were based upon skulls, simply because the rest of the skeleton was seldom preserved, and although the same is still largely true today, there are some species that do have some adequate post-cranial material. We will proceed by working our way down our list of skeletal features, noting the condition in the cynodonts and scoring them as being in the reptilian or in the mammalian conditions. If the condition is intermediate between the two, they will be so scored. Our comparison will be largely based on two genera, both from the Middle Triassic of Argentina: *Probelesodon*, because the skull is rather well known (nine skulls have been found, some of them being rather well preserved), and *Massetognathus*, for which we have some fairly complete post-cranial material. When we have completed our survey we will consider a third genus, *Probainognathus*, also from the Middle Triassic of Argentina, because it has a particularly interesting jaw joint.

Adding up the scores we have: Reptilian condition 5, Mammalian condition 5, Intermediate condition 4. Little wonder that we sometimes have difficulty in deciding whether these fossils are reptiles or mammals, and there is at least one case where a paleontologist first identified a particular species as being a reptile, only to change his mind a few years later and classify it as a mammal.

How should we classify these problematic fossils? Dr. W. G. Kühne, and some other paleontologists, believe that it is futile to try and draw the line between reptiles and mammals. Dr. L. Van Valen has suggested that the cynodonts and their relatives should actually be classified as mammals. Others prefer to draw the line on the condition of the jaw joint: if the joint is formed between the articular and the quadrate, the specimen is classified as reptilian, but it is classified as a mammal if the joint is formed between the dentary and the squamosal. But, as we saw in Chapter 6, there are some fossils where both types of jaw joint are present. The Argentinian cynodont *Probainognathus* is one such example, and if we examined its jaw joint we would see that there is a contact between the articular and the quadrate bone (reptilian condition) and between the dentary and the squamosal bone (mammalian condition). Furthermore, there

Checklist of Reptilian and Mammalian features of Cynodonts

	Reptilian condition	*Mammalian condition*
SKULL	1.	Teeth specialized for different functions.
	2. Cheek teeth with cusps but these are not complex. (Intermediate)	
	3. Lower jaw comprises several bones (note, however, that the dentary is the largest, the other bones being small).	
	4. Jaw joint formed between articular bone and quadrate bone (the quadrate bone, however, is very small).	
	5. Jaw joint formed between a hollow in the lower jaw and a flat surface in the skull. (Intermediate)	
	6.	Lower jaw with prominent coronoid process.
	7. Small cranium.	
	8.	Double condyle at back of skull for neck articulation.
POST-CRANIAL SKELETON	9.	Axis with odontoid process.
	10. Ribs in neck region.	
	11. Prominent ribs confined to chest region, but there are short ribs in front of the pelvis. (Intermediate)	
	12. Legs not splayed, but not placed vertically beneath the body either. (Intermediate)	
	13. Number of bones in fingers and toes exceeds 2,3,3,3.	
	14.	Ilium slopes forwards.

is a hollow groove on the skull which could be described as a glenoid, into which the lower jaw fits, and this is a mammalian feature. The quadrate bone is small, and is not firmly united to the rest of the skull. As a consequence, this bone has been lost in some of the specimens of *Probainognathus*. The fate of the quadrate and articular bones during the evolution of mammals is an intriguing story, and one that warrants a brief digression. Some readers, however, may feel that they have had enough anatomy at this point, and may skip the next two paragraphs without losing the thread of the story.

Mammals have a chain of three small bones, called the ear ossicles, which transmit vibrations from the eardrum to the receptor organ deep inside the skull. Starting from the eardrum these bones are named the hammer (malleus), the anvil (incus), and the stirrup (stapes), because of their shape. Reptiles, in contrast, have only one ear bone, the stapes. By studying the embryonic development of mammals it was found that the outer bone, the malleus, actually started off as part of the lower jaw, and that this articulated with the incus, which is part of the skull. The embryonic jaw joint is therefore formed between the malleus and the incus, and this led to the conclusion that they represented the articular and quadrate bones of the reptilian ancestor.

If we examined *Probainognathus*, we would see that the quadrate is already partially free from the skull. Furthermore, the stapes, which has been found in an incomplete condition in some of the specimens, lies in line with the quadrate, and it is believed that it actually articulated with it, as it is known to do in other cynodonts. In *Probainognathus*, then, we see an early stage in the loss of the reptilian articular-quadrate jaw joint. It is not difficult to visualize the next step, when the articular and quadrate bones are freed to link up with the stapes to form the mammalian ear ossicles. Dr. Gish's contention that nobody has been able to explain how the transitional fossil could have chewed its food while its jaw was being unhinged and reconnected is without foundation.

To return to our question of how to classify these problematic transitional fossils. As we saw in Chapter 6, we cannot use the criterion of the jaw joint because of problem fossils like *Probainognathus*. Most paleontologists overcome the difficulty by considering the jaw joint *and* the complexity of the teeth. To be classified as a mammal, a fossil has to have a dentary-squamosal joint *and* have accessory cusps on the cheek teeth. The fact that paleontologists have such problems in classifying these fossils underscores the point that they really are inter-mediate between reptiles and mammals, and Dr. A. S. Romer concluded his study on *Probainognathus* by saying that "I see nothing in the structural pattern of such a form as *Probainogna-thus* which need debar it from a position directly antecedent to a primitive mammal." Dr. Morris, I suspect, is a little uneasy about cynodonts. He comments: "The fact that it may be diffi-cult to tell, for example, whether a certain fossil was a reptile or a mammal does not mean at all that it was transitional be-tween the two in an evolutionary sense." But this statement makes no sense to me.

We have no transitional forms alive today, but there are cer-tain mammals, collectively known as monotremes, which have retained the ancestral reptilian condition of laying eggs. There are only two species, both living in Australia: the spiny ant-eater or echidna, and the duckbilled platypus. Both animals are highly specialized; the echidna is covered with sharp spines, reminiscent of the European hedgehog, and the platypus has a long duck-like bill and webbed feet. Both are toothless, but Dr. Gish finds it difficult to rationalize such a specialized or advanced feature in mammals that are supposedly the "most primitive." His difficulty stems from his lack of understanding that evolution does not proceed on all fronts at once, and this example of the retention of a primitive feature (egg-laying) is another instance of what we call mosaic evolution (that is, when organisms possess a mosaic of primitive and advanced features). Intermediate fossils, like *Archaeopteryx* and the cynodonts, are other examples of mosaic evolution. To give an analogy, the 1976 Chrysler I drive possesses a mosaic of primitive and ad-vanced features. Its "slant-six" engine, for example, is essen-tially identical to that of cars built more than twenty years ago, whereas its electronic ignition system is a recent innovation.

It is interesting to note that the monotremes, in addition to laying eggs, have a lower metabolic rate than the other mammals. Metabolic rate is a measure of the amount of heat generated in the cells of the body by their internal chemical processes. Reptiles have a much lower metabolic rate than mammals, and it is their high metabolic rate that enables mammals and birds to maintain high and constant body temperatures. It has been shown that the metabolic rate of the platypus, which is about 30 percent higher than that of the echidna, is 8 percent less than that of the pouched mammals (marsupials), and 35 percent lower than that of the other mammals (placental mammals). In this regard the physiology of the platypus and the echidna is intermediate between that of reptiles and that of the higher mammals (placental mammals). I interpret this as additional evidence for the reptilian ancestry of the mammals.

The fossils we have examined in this chapter bridge the gap between reptiles and mammals so perfectly that we have considerable difficulty in assigning many of them to their appropriate group. A similar situation was seen in the previous chapter regarding the transition from reptiles to birds. These intermediate fossils falsify the creationists' claim that transitional fossils linking major groups do not exist, and provide compelling evidence for evolution. We will examine some more examples of transitional fossils in the next chapter.

TWELVE
Horses, Hooves, and Embryos

 MOST MUSEUMS WITH DISPLAYS OF FOSSILS have a series of skeletons or parts of skeletons depicting the stages in the evolution of the horse. The skeletons range in size from the dog-sized *Hyracotherium*, previously called *Eohippus*, or the "dawn horse," to the modern horse *Equus caballus*. If a visitor takes the time to look at the skeletons and read the labels, he learns that the major changes that have occurred include a reduction in the number of toes, an increase in body size, an increase in the height and complexity of the cheek teeth, and an increase in the length of the gap (diastema) between the front teeth and the cheek teeth. The evolutionary history of the horse is probably the most widely cited example of a fossil series that supports the theory of evolution, but the creationists tell us that transitional fossils between the major types in the series are missing. We will examine the fossil evidence and see how wrong they are. We will also look at some embryological evidence which provides independent corroboration of the evolutionary conclusions drawn from the fossils.

The first point that has to be made, one that is seldom recognized, is that the various fossils in the series are not horses; only the modern horse, *Equus caballus*, is a real horse. Nobody would make the mistake of calling a zebra a horse, or an ass a horse, even though they are all members of the same genus, *Equus*. We should therefore not make the mistake of calling any of the fossils in the series horses. When we look at this evolutionary sequence, then, we are not looking at modifications within the "horse kind," as creationists might refer to them, but at major evolutionary changes within a particular group of hoofed mammals. The second point is that the skeletons usually displayed in a museum exhibit represent only the main stages, and the story which is told is therefore an oversimplification of the true situation. There are all manner of side branches and interrelationships that are not depicted, and any particular stage in the series is not necessarily considered to be the direct ancestor of the next. This evolutionary series is comparable to representing the evolution of the automobile by a horseless carriage, a 1920 racing car, and the latest model from Ford.

In spite of these qualifications, the series gives an excellent documentation of the evolutionary changes taking place within a closely related group of animals. We will now look at the major stages and evolutionary trends in this series, and then focus upon one aspect: the evolution of the foot. The reason we will be giving the foot so much attention is that it has undergone extreme modification in the development of the complex springing action seen in the modern horse, and there is some interesting embryological evidence which supports our evolutionary conclusions.

The horse and its living relatives, the rhinoceros and the tapir, together with their extinct relatives, belong to a group called the odd-toed ungulates, or perissodactyls, to give them their scientific name (perisso = odd; dactylos = toe). They are so named because they have an odd number of toes; the rhinoceros and the tapir, for example, have three toes, while the horse has only one. (Actually the tapir, like *Hyracotherium*, has three toes in the hind foot but four in the forefoot.) We have already said that the ancestors of the modern horse are not really horses; the group name for these fossils, and for the modern horse and its living relatives, the zebra and the ass, is the family Equidae, or equids for short.

The earliest equid in the evolutionary series is the terrier-sized *Hyracotherium*, which lived during the Eocene Period, about sixty million years ago. *Hyracotherium* has three toes in the hind foot (the big toe, digit 1, and little toe, digit 5, have been lost), and four toes in the forefoot (digits 2, 3, 4, and 5).

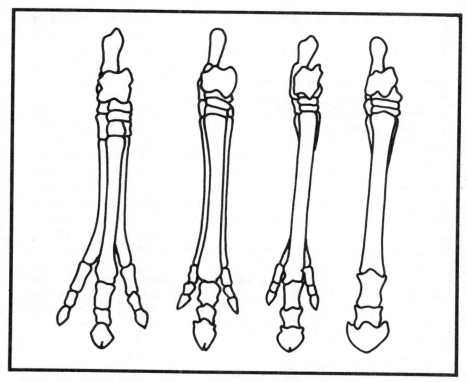

Comparisons of the hind legs of three horse ancestors and a modern horse. From left to right: Hyracotherium, Mesohippus, Merychippus, *and* Equus *(modern horse).*

The jaw and the facial region are not deep as they are in the horse, and this is largely because the teeth were low-crowned and therefore did not require a deep jaw to accommodate them. The cheek teeth lack the complex cusp pattern seen in later equids, but the molars already had an oblong grinding surface, and there was a small but distinct diastema (gap) between the premolars and the incisors.

The next stage is represented by *Mesohippus*, which is about the size of a German Shepherd dog and which lived about thirty million years ago, during the Oligocene Period. Here the front and the hind legs both have three digits (digits 2,3, and 4) and the lower leg region is relatively longer than it is in *Hyracotherium*. The cheek teeth are still low-crowned, with relatively short roots, but most of the premolars have an oblong grinding surface, like the molars, and the cusp pattern is more complex in both. The diastema is larger than in *Hyracotherium*.

There is no chance that either of these two equids would be mistaken for horses, because their skeletons obviously differ from that of a modern horse. From their low-crowned teeth we surmise that they were browsers, that is, that they fed upon relatively soft vegetation such as leaves, as do present-day tapirs and a host of even-toed ungulates, like antelopes. Leaves are far less abrasive than grasses; animals that feed on grass, called grazers, wear their teeth down so rapidly that they have to have either high-crowned cheek teeth, like horses, or cheek teeth that grow continously, like sheep. It is no mere coincidence that the next major stage in the evolution of equids, as represented by forms with high-crowned teeth, such as *Merychippus*, appeared during the Miocene, some twenty million years ago, at a time when grasses appeared and spread rapidly.

Merychippus, whose height ranged from that of a small pony to that of a regular horse, has most of the features that we associate with the modern horse. The teeth are high-crowned; consequently the lower jaws and the facial region are deepened to accommodate them. As the crowns were worn down during the life of the individual, they were extruded from their sockets, just as in a modern horse. At the end of a horse's life the crowns of its cheek teeth may be only about a centimeter long, whereas they started off at more than ten times this length. Premolars and molars are both rectangular in *Merychippus*, as in the modern horse, and the cusp pattern is similarly complex. There is also a large diastema, just as in the horse. The feature that clearly sets *Merychippus* apart from the horse is its three-toed foot, and there is a wide range of variation in the relative lengths of the side toes among the various species that are placed within the genus. In some, the side toes are quite long, perhaps 2/3 of the length of the middle toe, but, because the metapodial bones (equivalent to the bones in the palm of our hand and in the sole of our foot) to which each is attached is a little shorter than that of the middle toe, they stop a long way short of the middle toe. In other species the side toes are relatively much shorter, and probably did not contact the ground except when the animal was galloping. To understand how the side toes may have functioned, we need to understand how the foot operates in a modern horse.

The horse has only one toe, which terminates in a hoof, and when this is placed upon the ground and the body weight is transferred to it, the hoof is extended, like the terminal bone in our index finger when we are pressing down on a pen. The horse's hoof is prevented from extending too far by several ligaments and tendons of muscles that run along the back of the leg. These ligaments and tendons are called the suspensory apparatus, and are especially thick and elastic, functioning like a spring. When the hoof strikes the ground during the gallop, it is extended such a long way forward that the suspensory apparatus is tightly stretched, and therefore stores a considerable amount of energy. As the weight is being taken off the hoof it snaps back under the springing action of the suspensory ligament, and this contributes a driving force to the foot. This springing action of the hoof is an integral part of the locomotory apparatus of the horse, and it was probably present at an early stage of its evolution at the stage represented by *Merychippus*. When *Merychippus* was running, it is likely that the side toes functioned as braces, assisting the checking action of the suspensory apparatus as the middle toe was extended by the body weight. The fact that the middle toe, and its corresponding metapodial bone, is considerably better developed than the side toes shows that it bore most of the weight of the body.

We therefore see a reduction in the weight-bearing role of the side toes during the evolution of the equids, culminating in the one-toed foot of the modern horse. What is particularly interesting to us here is that we can see all shades of gray, from equids like *Mesohippus*, where the side toes and the middle toe are essentially all the same length, through to some species of *Merychippus*, where the side toes are so small that they probably only made contact with the ground when the animal was galloping. But this is contrary to what Dr. Gish tells us. Citing from papers that are more than thirty years old, he claims that the transitional stages between the major types of equids are missing. As far as the evolution of the foot is concerned, the only stage which appears to be missing is the final stage, from an equid like *Merychippus*, with two small side toes, to the modern horse, which has no side toes at all. However, when we look closely at the anatomy of the modern horse, and when we consider its embryonic development, we can see that there is no gap at all.

Viewed from the front, the foot and lower part of the leg of a horse look somewhat different from those of *Merychippus*, and this is because there is no evidence of side toes. However, the metapodial bone which supports the middle toe, called the cannon bone, is just like the middle metapodial bone of *Merychippus*, and if we view it from the back, the resemblance to *Merychippus* is even greater. Running along the posterior surface of the cannon bone are a pair of bones, one on either side, which are called splint bones. The splint bones occupy a similar position to the metapodial bones that support the side toes in *Merychippus,* and there is no question that they are the metapodial bones of the ancestral side toes (toes 2 and 4). This is compelling evidence for descent with modification, and it is very difficult to rationalize the retention of splint bones in the horse other than by evolution.

Is there any evidence of side toes in modern horses? If there were, it would strengthen our case for descent with modification. "Since . . . Gegenbaur suggested that the presence of extra digits might in some cases be due to the existence of latent 'germs' in the embryo, many have looked for vestiges of the phalanges of the second and fourth digits [in the horse]. But the quest . . . has hitherto been in vain," wrote the anatomist J. C. Ewart in 1894. He goes on to tell us that he started his investigation without any hope of finding rudiments of the side toes in the embryo, or any vestiges in the adult horse, but that he did in fact find them. Dr. Ewart studied horse embryos ranging in size from only 2 cm in length to about 88 cm, and he noticed that small buds appeared, one beneath each of the developing splint bones, at a fairly early stage. These buds were quite small, only 3 mm long in the 2.5 cm embryo and no more than 5 mm long even in a 35 cm-long embryo. With patience and great care he was able to dissect the buds and study their internal structure. He was amazed to find that the buds were rudimentary side toes, complete with tiny caps at the end which he believed represented hooves. The side toes were at their most advanced stage of development in the 35 cm-long embryo, where it was possible to distinguish three individual elements corresponding to the three bones in the side toes of *Merychippus.* After the 35 cm-embryo stage the individual elements lost their separate identities, which was the ultimate fate of the whole bud.

Dr. Ewart correctly surmised that the embryonic development of side toes in the horse is evidence that horses have inherited the genetic capacity to form side toes from their three-toed ancestors. The structural genes that control the formation of side toes get "switched off" by regulator genes during embryonic development; consequently the side toes are lost prior to birth. Sometimes, however, something goes wrong with these regulator genes, and the occasional foal is born which has well developed side toes. Such an instance was reported by Dr. John Struthers in 1893, and there have doubtless been other cases since.

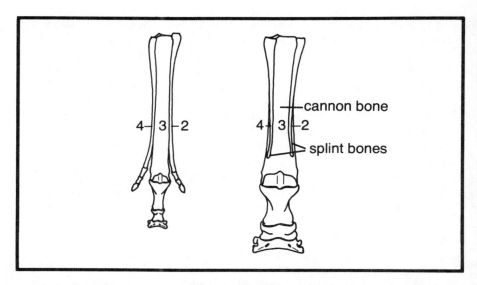

Comparison of the lower left leg of Merychippus *(left) and the modern horse, shown from the back. Notice that the modern horse has remnants of metapodial bones 2 and 4 (splint bones).*

There is no question, then, that horses have retained the genetic capacity to form side toes, just as the chicken has retained the capacity to form teeth (Chapter 10), and this is evidence that they have evolved from a three-toed ancestor. The alternative proposition, that the modern horse was created, is difficult to reconcile with the evidence. Why would the Creator give the horse the capability of developing side toes, a capability always exercised during embryonic development but rarely manifest after birth? The side toes that form during early embryology cannot be dismissed as developmental accidents that have nothing to do with real toes, because they each comprise three elements, which perfectly match up with the three bones seen in the side toes of equids like *Merychippus.* Perhaps creationists might argue that the differences between a three-toed and a one-toed horse are not significant, being merely variations within the horse "kind." This is not a valid argument, however, because three-toed animals like *Merychippus* are *not* horses; they are much further removed from a modern horse than a modern zebra or an ass, neither of which would be described as a horse, even by the most uncritical observer. As far as I can see, creationists are faced with a serious problem over the question of horses; and perhaps this is why neither Dr. Morris nor Dr. Gish has very much to say on the subject.

THIRTEEN

Newts, Lungfishes, and Sea Dragons

 ACCORDING TO EVOLUTIONISTS, the first terrestrial vertebrates, the amphibians, evolved from fishes, and these, in turn, gave rise to the reptiles. The creationists deny that there is any evidence to support either of these transitions, but we will show that they are wrong. We will also present evidence for the transition of a group of marine reptiles, the plesiosaurs, from terrestrial ancestors.

One of the things I liked about spring when I was a boy was going to collect frog spawn. Motley gangs of us, equipped with empty jam jars and improvised nets, would set off for local ponds and streams in the neighborhood to see how much of the gelatinous egg masses we could collect. Sometimes one of us would get lucky and find a long strand of jelly instead of the usual clump: the highly prized eggs of the newt.

Much of the spawn we collected found its way down the backs of dresses, into birdbaths and ornamental ponds, and sometimes through letter boxes, but some was kept long enough to hatch into tadpoles. Occasionally we would be able to persuade our mothers to let us keep the tadpoles long enough to see them develop into small frogs, but I cannot recall ever rearing any newts to this stage of development. Newts, being less common than frogs, were highly prized, and were usually taken as adults rather than as eggs. These small amphibians, not much longer than one's finger, look like miniature dragons with their frilled backs and long tails, especially in the breeding season when their bellies turn fiery orange.

Newts belong to a group of amphibians called the urodeles, meaning visible tail, while frogs and toads belong to a separate group called the anurans, which means tailless. Some of the urodeles, like the salamander, spend most of their time on land, breathing through lungs as well as through their skin, while others, like the mud-puppy, spend all of their time in the water and even have external gills for breathing. Many other urodeles possess external gills, too, and some of these have such small legs (in one called the siren the hind legs are missing altogether) that they look somewhat like fishes. The species with external gills look very much like the larvae of the land-living species, and by administering hormones they can often be made to lose their gills, changing into adult-looking amphibians. We therefore see all shades of gray among the amphibians, from fully aquatic animals, reminiscent of fishes, to fully terrestrial ones that remind us of lizards. Furthermore, most amphibians pass through a tadpole stage during their life history, a stage that looks very much like a fish.

The word amphibian actually means both lives, reflecting the readiness with which most of them can move between land and water, and they provide us with ample examples of how the transition from water to land can be made. We are not suggesting that modern amphibians represent an intermediate stage between fishes and reptiles — they are all too specialized for that — but they do belong to the same class of vertebrates that made the transition, way back in the Devonian Period. This transition was from an air-breathing fish to an amphibian.

In the 1830s an unusual fish was caught in a river somewhere in South America, and subsequently sent to Germany for scientific study. This was a fairly large fish, about one meter long, with a slender body gently tapering to a symmetrical tail. The narrow fin that fringes the tail starts about halfway down the length of its back, and continues on the underside of the body as far forward as the paired hind fins. Unlike the paired fins of any other living fishes, which have the form of a fan of rays radiating out from a small base, these fins are long and fleshy. The same is true of the forefins, and in life these fins are used for "walking" on the river bed. Even more peculiar is the way that this fish comes to the surface to breathe air; to do this it is equipped both with internal nostrils and with lungs, structures that are usually found only in land animals (tetrapods). The gills, and the blood vessels that supply them, are relatively less well developed than in other fishes, and the heart, in having two auricles instead of one as in other fishes, is more like that of a land animal. Little wonder that the German scientists who studied the first specimen, which was named *Lepidosiren* (meaning scaled siren), concluded that it was an amphibian.

The second example of these unusual animals, which are today called lungfishes, was sent to the British anatomist Richard Owen. He made a detailed study of the material, which was from Africa, and named it *Protopterus*. He also gave a detailed account of *Lepidosiren* (published in 1841) in which he discussed the problem of to which vertebrate group these animals should be referred. He was torn between classifying them as fish or as amphibians but in the end he concluded that they should be classed as fish. Owen's word carried a great deal of weight among scientists, but this did not stop some of them from claiming that the lungfishes should really be classified as amphibians. A consensus emerged from the discussions — lungfishes really should be classified as fish rather than as amphibians, but they were the most amphibian-like of all fishes. Owen underscored their transitional position between fishes and tetrapods in an entry in the Encyclopaedia Britannica for 1859, the year of the publication of Darwin's *Origin of Species*: "In this structure [anatomy of the vertebral column] the old carboniferous reptile [*Apateon*] resembled the existing *Lepidosiren*, and affords further grounds for regarding that remarkable existing animal as one which obliterates the line of demarcation between the fishes and the reptiles." Owen, we should point out, could not be described as an evolutionist; indeed, it was he who had supplied ammunition to Bishop Wilberforce in the famous Oxford debate with T. H. Huxley, a point to which we shall return later.

A third lungfish has since been found, *Neoceratodus*, from Australia, and this has the most highly developed fore and hind limbs of the three. These muscular limbs are used for walking on the bottom of rivers where they live, and they can walk both forwards and backwards. The rivers these lungfishes live in are subject to periodic stagnation, sometimes even to complete drying up, and when this happens the fishes dig into the mud, leaving a small breathing-hole, and remain there in a state of inactivity until the rains come. Lungfishes are found in the fossil record, dating back to the Devonian Period, and mud burrows similar to those made by living lungfishes have also been found.

The lungfishes, together with the coelacanths and some fossils called the rhipidistians, are collectively called Sarcopterygii, or fleshy-finned fishes. (The coelacanths and rhipidistians are collectively called crossopterygians.) The only surviving sarcopterygians are the three species of lungfishes, and one species of coelacanth, which was discovered off the African coast in 1938. The sarcopterygians are of particular interest to us because they are the group that we believe were ancestral to the amphibians. As we saw earlier, the idea that sarcopterygian fishes were related to amphibians dates back to the time before Darwin published his theory of evolution when Owen, who was not an evolutionist, drew attention to their intermediate nature. But Dr. Gish implies that it was *evolutionists* who first perceived the connection when he tells us the evolutionists, lacking a candidate intermediate between fish and amphibians, searched the various fish groups, finally selecting a subgroup of the sarcopterygians, the crossopterygians, as the most likely ancestor.

If we look at the internal skeleton of the paired limbs of sarcopterygian fishes, especially certain rhipidistians like *Eusthenopteron* and *Sterropterygion*, both Devonian in age, we see a number of similarities with those of the earliest amphibians such as *Ichthyostega*, which appeared at the close of the Devonian Period. If we compare forelimbs we see the following similarities: there is a fairly complex humerus (upper arm bone) which has a pair of distal (lower) facets for articulation with the two bones of the forearm, the radius and the ulna; the humerus is twisted, such that the axis which passes through the distal facets is set off at an angle to the long axis of the head of the humerus (which attaches to the shoulder). This twist in the humerus, which is found in land animals but not usually in fishes, is correlated with the positioning of the limbs so that they make contact with the ground. The hind limb of *Eusthenopteron* is not as well developed as the forelimb, but the main elements can be discerned: the femur, tibia, and fibula, which correspond to the amphibian pattern.

The similarities between *Eusthenopteron* and *Ichthyostega* extend beyond the forelimbs. Their skulls are both similar, having the same basic pattern of bones which form the roof of the skull, the palate, and the lower jaws. Furthermore, there is evidence of shallow grooves on the skull roof of many of the early amphibians which correspond to those of fishes, and which are

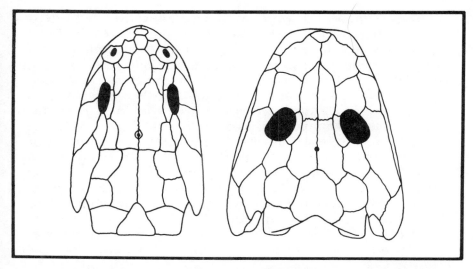

Comparison of the skulls of the fossil fish Eusthenopteron *(left) and the fossil amphibian* Ichthyostega. *Although some bones have disappeared in the amphibian, and some have become fused, the basic patterns are similar.*

part of the lateral-line system. This is a sensory system that allows the animal to detect water-borne vibrations. *Eusthenopteron* and its allies also have internal nostrils (or nares) as well as external ones, a feature found in land animals and not in any other fishes except sarcopterygians. Their teeth are also very similar to those of primitive amphibians, both being sculptured by deep ridges, and when the teeth are examined in cross-section they look something like a rolled tube of corrugated cardboard. But perhaps the most striking feature of the rhipid-istians and their allies is that their braincase comprises two parts, anterior and posterior, between which there seems to have been a certain degree of mobility. No other vertebrates have a two-part cranium, none, that is, except for the ichthyo-stegid amphibians. The sharing of such an unusual feature is compelling evidence for their being related.

Living land animals have backbones comprising individual bones, the vertebrae, which are essentially spool-shaped disks of bone attached to various spines and processes. The vertebrae of *Ichthyostega* and its allies, however, did not have a complete bony disk, having instead two crescents of bone, a lower and larger one called the intercentrum, and an upper one called the pleurocentrum. The two parts were not fixed together, and a similar vertebral structure is found in *Eusthenopteron*.

These similarities, which we have only considered briefly, are very persuasive evidence that the amphibians evolved from crossopterygian fishes, but the creationists remain unimpressed. Dr. Gish complains that not a single transitional fossil has been found which shows an intermediate stage between a crossopterygian fin and an ichthyostegid foot. He then goes on to ask why a fish like *Eusthenopteron* has been chosen as a possible ancestor for the amphibians, concluding that there was nothing better. Dr. Gish is correct when he says that there are no fossils intermediate between crossopterygian fishes and ichthyostegid amphibians, but his diagram depicting the width of the gap between the two would have been narrower had he figured *Ichthyostega* instead of *Eryops* as "an ichthyostegid amphibian." I am sure that this was an unintentional mistake on Dr. Gish's part (he did not repeat the error at a recent presentation he gave), but the fact remains that *Eryops* is far less fish-like than *Ichthyostega*.

Dr. Gish goes on to attack the textbook scenario of how amphibians arose during the Devonian. According to this widely held view, the amphibians evolved in response to periodic drying up of the fresh waters that their sarcopterygian ancestors were living in. If this scenario is right, argues Dr. Gish with sound logic, why was the Devonian not a period of mass extinction of sarcopterygian fishes and other freshwater animals? Dr. Gish is not the only person to criticize this scenario, and Dr. K. S. Thompson, among others, has presented evidence that most of the sarcopterygian fishes lived in, or near, the sea. He argues that the selection pressure for the evolution of the amphibians was not periodic drought but the availability of abundant food on land. We should also point out that if the drought scenario were true, we should expect to see the survival of sarcopterygians rather than their extinction, because, as we saw earlier for the lungfishes, air-breathing fishes have an obvious advantage during periods of stagnation and drought.

The last point that Dr. Gish raises pertains to details of the structure of the vertebrae of fishes and amphibians. As we have seen, complete bony disks are not present in crossopterygian fishes, or in amphibians like *Ichthyostega*, this part of the vertebra being represented by two crescentic wedges of bone, the intercentrum below and the pleurocentrum above. This pattern is called rhachitomous, and there are three other patterns found in amphibians: lepospondylous, embolomerous, and stereospondylous. In the first type, found in numerous small amphibians (mostly Carboniferous and Permian in age, not Devonian),

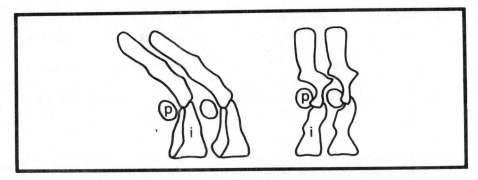

Comparison of the similarity of vertebral structures in the fossil fish Eusthenopteron *(left) and the fossil amphibian* Ichthyostega. *In each case, two vertebrae are shown, from the right side (p=pleurocentrum; i= intercentrum).*

the vertebrae are spool-shaped disks and do not comprise two separate pieces as in all the other amphibian vertebrae. These are often called "husk vertebrae," while the others are called "arch vertebrae." In the second type, with the embolomerous vertebrae, the intercentrum and the pleurocentrum are both present, but, in contrast to the rhachitomous vertebrae of *Ichthyostega* and its allies, the two parts are approximately of equal size. The third group, with the stereospondylous vertebrae, have either lost the pleurocentrum altogether, or else it is only very small.

Dr. Gish does not discuss these various types of vertebrae, but he makes the point that three other amphibian orders are found in rocks of early Carboniferous age (that is, the period immediately following the Devonian, when *Ichthyostega* appeared), and mentions two of them, the Orders Aistopoda and Nectridea. He then tells us that all three "possessed the 'more primitive' lepospondylous, or 'husk' type, vertebrae," and asks how the arch type of vertebra can be used to link crossopterygians and amphibians.

Just why he thinks the lepospondylous vertebra is more primitive is not clear, because the amphibians which possessed it all appear in the fossil record *after* the ichthyostegids. Since he has the primitive and the more specialized types of amphibian vertebrae switched around, it is hardly surprising that his discussion of amphibians should be confused. The fact that the arch vertebrae are found in the oldest of amphibians, that the three patterns of arch vertebrae are all variations on a common theme, and that one of them, the rhachitomous pattern,

is seen in rhipidistian fishes, is compelling evidence that amphibians evolved from fishes. We will now move on to the
transition from amphibians to reptiles.

As we saw in Chapter 8, we often have great difficulty in
assigning fossils to their appropriate major groups when we
are close to the border between them, and we spent Chapters
10 and 11 discussing two such transitions, from reptiles to birds
and from reptiles to mammals. We now have the same problem
in the transition from amphibians to reptiles, and this is
compounded by the fact that most of the features used to distinguish between the two groups pertain to features that are
not preserved in fossils. The major difference between amphibians and reptiles is that reptiles have attained complete inde-
pendence from the water, and the features that have made this
possible include: laying shelled eggs, having skin that is essentially impermeable to water, and having well-developed
lungs. These features are not recorded in the fossil record, and,
as Drs. Gish and Morris point out, the skeletons of ancient
amphibians and reptiles are so similar that there is little way
of distinguishing between them. This, of course, is precisely
what an evolutionist would expect to find, and parallels the
situation we have already seen in the other transitional series.

Do the physiological differences between amphibians and
reptiles, those features that are not preserved in the fossil record,
represent an insurmountable gap between the two groups?
Amphibians do not lay shelled eggs, like those of reptiles, but
many species lay eggs that have a tough capsule around them,
as in the case of the microhylids, a group of small tree-frogs
found in the tropics. The embryos develop all the way into
tiny frogs inside the capsule, the tadpole stage having been
eliminated. The egg capsules of some frogs, including species
of the genus *Discodeles*, a large frog found in Australasia, is so
tough that the young frogs are equipped with a spike on the
tip of their snouts to puncture the capsule. A similar "egg tooth"
is found in some reptiles and in birds.

Laying encapsulated eggs is not the only strategy that has
evolved in amphibians that enables them to breed on land
without the necessity of returning to water. Many species make
a foam nest around their eggs, others lay their eggs in damp
places (under stones and beneath the bark of fallen trees), and
others, like the midwife toad (*Alytes obstetricans*), carry their eggs
around with them, occasionally moistening them with dew.
One of the most extreme strategies is seen in tree-frogs belonging to the genus *Pipa*, where the eggs develop in fluid-filled

pouches beneath the skin on the back of the female. Some amphibians, including the legless caecilians (worm-like animals that burrow in the soil) and frogs like the tailed frog (*Ascaphus truei*), fertilize their eggs internally, like reptiles, birds, and mammals, and the embryos actually develop inside the female's body. This brief survey of reproductive strategies shows that the gap between amphibians and reptiles is not as wide as creationists would have us believe. What about the two other aspects we mentioned: skin permeability and lungs?

While some amphibians, like toads, have skin that is relatively dry and less permeable to water than that of other amphibians, none of them have the tough waterproof skin that characterizes reptiles. But that does not mean that amphibians are unable to live in the drier regions of the world, and some of them actually live in deserts. Amphibians have been able to achieve this by modifications in their lifestyles rather than in modifications of their skin. Most amphibians only come out at night, thereby avoiding the drying rays of the sun, having spent the day in burrows, beneath stones, under logs, or in some similar sheltered spot. Frogs living in the Australian desert, for example, have been seen huddling together in rock crevices during the day and this behavior drastically reduces the amount of water lost to the atmosphere. Frogs living in these dry regions store water in their bladders, a fact that is known to thirsty Australian bushmen who seek them out as a source of water.

As far as breathing is concerned, the skin is an important respiratory surface, but most amphibians possess lungs too. While these are fairly simple structures in most species, usually lacking the convolutions that are seen in the higher vertebrates, convoluted lungs are found in some species, especially those that are terrestrial. We therefore see an approach towards the reptilian respiratory system in some amphibians, though they probably all rely heavily on breathing through the skin.

These three physiological examples — reproduction, water retention, and respiration — show that the distinctions between amphibians and reptiles, while significant, are not insurmountable, and the many and obvious similarities between the two groups are compelling evidence for their evolutionary relationship. Dr. Gish, of course, denies this relationship, and one of his major objections relates to timing.

The fossil amphibian *Seymouria*, Dr. Gish points out, has a skeleton so much like that of reptiles (paleontologists once classified it as a reptile) that it has long been held as a perfect transitional form between the two groups. The only problem is, though, that *Seymouria* is Permian in age, whereas the first reptiles appeared in the preceding Carboniferous Period. Dr. Gish has a valid argument, but he misses the point that *Seymouria* is only one member of a group of amphibians, a group that first appeared in the Carboniferous Period.

For many years now I have been interested in a group of marine reptiles, called ichthyosaurs (meaning fish-lizards), that were contemporaneous with the dinosaurs. They were so highly specialized for living in the sea that they looked superficially like fishes, and this is reflected in their name. Ichthyosaurs first appeared in the fossil record in the early Triassic, were particularly abundant during the early part of the following Jurassic Period, and survived into the succeeding Cretaceous Period. Large numbers of specimens have been collected from Jurassic localities in England and Germany, and some of the German ones are in such good condition that their body outlines have been preserved as a carbonaceous film. Like fishes, they had a dorsal fin on the back, they had a large crescent-shaped caudal fin which was partly supported by a downward bend in the vertebral column, and their fore and hind limbs were fins rather than arms and legs. The primitive five-fingered pattern (called primitive, or ancestral, because the earliest tetrapods had five fingers) that we see in our own hands and feet can usually not be seen because the fingers, which frequently exceed five in number, are pressed tightly together and there are far more than the usual number of bones in each finger.

If we compare *Mixosaurus*, the best-known of the earliest ichthyosaurs, with one of the Jurassic or Cretaceous ones, we can see that several of the typically ichthyosaurian features had not yet evolved. There was no sharp bend in the vertebral column, so that the caudal fin was neither crescent-shaped nor large, and the limbs, while obviously modified as fins, still have five distinct fingers, and these are without the large number of individual bones seen in the later forms. But *Mixosaurus* was already recognizable as an ichthyosaur; where, then, did the ichthyosaurs come from? Do we have any fossils intermediate between *Mixosaurus* and an unspecialized land reptile?

When Professor Robert Carroll of McGill University, a paleontologist who is particularly interested in the relationships among reptiles, last asked me that question, I believe I suggested that ichthyosaurs had just dropped out of the sky. The embarrassing fact is that we have not yet found the ancestor of the ichthyosaurs. This has not prevented paleontologists from speculating, though, and most reptilian groups, at one time or another, have been proposed as possible ichthyosaur ancestors. Why have we not found the ancestor? Creationists would argue that the reason is simply that it never existed, but I prefer to believe that the ancestor did exist, and that we have just not found it yet. If that sounds like wishful thinking, let me respond by recounting a story about a similar group of reptiles, the plesiosaurs, which also happened to have lived in the sea during the age of dinosaurs.

Not many people are familiar with ichthyosaurs, but plesiosaurs are quite well known; they are those reptiles with the long neck, long tail, and big paddles fore and aft — the ones that some people like to believe are alive and well and living in Loch Ness. (Not all plesiosaurs had long necks; some had short necks, with relatively large heads, but they all had the same basic skeletal structure.) One of the distinctive features of the plesiosaur skeleton is the large size of the shoulder and pelvic girdles. These are like large plates which all but enclosed the abdominal region, providing a large attachment area for the muscles that operated the paddles.

Plesiosaurs appear in the fossil record at the beginning of the Jurassic Period, and, like their distant relatives the ichthyosaurs, they were already highly specialized from the very beginning. Where did they come from? This is not a difficult question to answer because there is a group of marine reptiles, called nothosaurs, which are Triassic in age and which have so many features in common with plesiosaurs that they are considered to be the group from which plesiosaurs evolved. Like plesiosaurs, they have well-developed shoulder and hip girdles, and although these are plate-like, they are not as well developed as in plesiosaurs. Similarly the fore and hind limbs, while relatively large, retain the basic pattern of legs rather than of paddles. There are many other similarities, including a streamlined body, a long neck and tail, and similar skull construction. Indeed, the two groups merge into one another in a way that is most satisfactory to an evolutionist, but a creationist would probably be unimpressed and would want to see a link between nothosaurs and unspecialized land reptiles. There is a group of Permian land reptiles, called the Eosuchia (meaning dawn crocodile), which are believed to be the ancestral stock from which the lizards evolved, and some paleontologists have suggested that they also gave rise to the nothosaurs. This idea, though, was not widely accepted, and there is, in any case, still a wide gap between these land reptiles and the marine nothosaurs.

Serendipity often plays a major role in the affairs of paleontologists, and it was very much by chance that one of the employees of an oil company that was operating in Madagascar a few years ago was interested enough in the fossil reptiles that were being unearthed to save them. The fossils, which are late Permian in age, found their way to Paris, and the French paleontologist J. Piveteau, after a preliminary examination, suggested that they might be ancestral to the plesiosaurs and their allies. Because of prior commitments Dr. Piveteau was unable to work on the material, and so it was that Professor Carroll was invited to study it, publishing his findings in 1981. Dr. Piveteau's suggestion was corroborated — here was a new type of reptile, named *Claudiosaurus*, which neatly filled the gap between the terrestrial Eosuchia and the aquatic Nothosauria. The fore and hind limbs are large, but are not modified for aquatic use and might have been used for walking on land or for swimming. The girdles to which they attached are large and plate-like, somewhat like those of a nothosaur, but are not so extensive. The skull is more modified than the eosuchian skull, but has not reached the nothosaur stage.

We are therefore able to trace an evolutionary pathway from fully terrestrial reptiles to the fully aquatic, and highly specialized plesiosaurs. We obviously do not have a complete series of finely graded steps, nor would we expect to have all the intermediates, given the vagaries of the fossil record, but we do have compelling evidence that the plesiosaurs did evolve from a terrestrial ancestor. One day we may be able to do the same for the ichthyosaurs.

The evidence we have considered in this chapter reinforces that of Chapters 10 and 11, firmly establishing the fact denied by the creationists that intermediate fossils linking major groups do exist. These intermediate fossils are uncommon, but, as we have seen in the case of the plesiosaurs, new ones are discovered from time to time.

Coming Down from the Trees: The Descent of Man

FOURTEEN

 WHEN GUY THE GORILLA DIED at the London zoo a few years ago, England was plunged almost into a state of mourning. Ever since his arrival at the zoo from France just after the war he had been the center of attraction, and his public image grew with his stature over the years that followed. Part of his appeal lay in his large size; he was indeed a most powerful animal, and spectators would often speculate upon the havoc he would wreak if he ever got loose. There is no question that gorillas are immensely strong, and their sullen expression engenders the impression of great ferocity, but when travelers to Africa stopped shooting them and began observing them, it was discovered that they were gentle and retiring creatures.

The feature that fascinated me most about Guy was his eyes — those inquiring eyes that would fix you and make you wonder just what he was thinking about. I doubt whether any of us can look at one of the great apes without entertaining some thoughts of our relationship to them. We have many features in common: hands with long, probing fingers, fingernails, facial expressions, an ability to stand up on the hind legs and walk, to mention just a few. And below the surface, under the fur and beneath the muscles, are a host of skeletal similarities, features that have been preserved as fossils.

Man's fossil record, while far from complete, adequately documents our descent from apelike ancestors through a series of transitional fossils. The creationists, of course, vehemently deny this, and variously dismiss these intermediate fossils as aberrant apes, undoubted humans, or outright frauds. In reviewing the evidence in this chapter, we shall see just how far from the truth the creationists have strayed.

A limestone cave in the Neander Valley, Germany was the burial place of one of the first fossils to be identified as human. The beetling brow, low crown, and thickness of the skull bone spoke of its primitiveness, as did its relatively small cranial capacity, estimated at little more than one liter by its original describer, Dr. D. Schaaffhausen. The timing of the discovery, just two years before the publication of Darwin's *Origin of Species*, could hardly have been more appropriate. This was Neanderthal Man, the archetypal caveman. Darwin, as we know, did not discuss man's ancestry in the *Origin of Species*, but this was the subject that was foremost in most people's minds when they read his revolutionary ideas. The topic is as hotly debated today as it was then, though usually for different reasons. Most discussion today is based upon the question of which particular ancestors gave rise to man, not whether man actually did evolve in the first place, and these discussions have arisen out of the exciting new fossils that have been found in the last decade or so. We will be largely concerned with documenting the evidence of man's descent from an ape-like ancestor, but before we begin, we have to know something about our skeletal anatomy, and how we compare with our more hirsute relatives.

We belong to an order of mammals called the primates, a group that took to living in the trees tens of millions of years ago. Most primates still live in the trees, but a few, notably ourselves and the gorillas, have taken up a life on the ground. Even though we no longer clamber about in the trees, our arboreal ancestry has left its stamp upon us, and the features we share with the other primates are adaptations to this way of life.

Tree-dwellers have to be able to move efficiently from branch to branch and different groups of primates have adopted different strategies for doing this. Monkeys walk along branches on all fours, jumping from branch to branch. The apes (the gibbon, orangutan, chimpanzee, and gorilla) swing from the branches by their forelimbs, which are relatively much longer than their hind limbs. The third group of primates, the prosimians, which includes such animals as lemurs, lorises, and bush-babies, are clingers and jumpers. They move about in the trees somewhat like squirrels, and are considered to be the most primitive of all the primates (that is, they are believed to retain more of the features of the group from which the primates evolved). Although the strategies for moving about in the trees vary, the basic requirements remain the same: the ability to judge distances, the ability to grasp branches, and, if movements are to be swift, an agile mind.

A prerequisite for mammals for judging distances is binocular vision, and all primates have their eyes facing forwards to achieve this end. It is exceedingly difficult to grasp objects, or make rapid changes of direction, without binocular vision. Just try threading a needle with one eye closed — but do not try riding a bicycle wearing an eye patch as I once did! We see evidence of binocular vision when we examine the skull of a primate because the orbits, or eye sockets, are directed forwards. Most, but not all, primates have nails instead of claws, the fingers and toes tend to be long, and the thumb and often the big toe, too, are opposable, enabling them to grasp and manipulate.

While the more primitive primates, such as the lorises and the lemurs, are not especially endowed with gray matter, the higher primates are. Not only are their brains relatively large, but that part which is involved with what we may loosely term mental ability, namely the cerebral cortex, is especially well developed. The higher primates — apes, man, and our ancestors, together with the cetaceans (whales and their allies) — have the largest of all brains relative to their body size. Brain size is usually estimated for extinct animals, and for living ones too, by measuring the capacity of the cranium (brain box). Because the mammalian brain fits quite snugly inside the cranium, the measure of the cranial capacity gives a good estimate of the actual volume of the brain. We should emphasize here that while the relative size of the brain is a general guide to the intellectual and behavioral capabilities of its owner, the relationship is only approximate. Reptiles, as we know, have relatively small brains, and correspondingly limited intellectual potentials; a 52 kg alligator, for example was found to have a brain volume of only 7 ml, compared with 325 ml for a 44 kg chimpanzee. In our own species, though, there is a surprisingly wide range of brain sizes, from about 900 ml to about 2200 ml, but there appears to be no relationship between brain size and intellectual potential.

The average size of the human brain is 1300 ml, compared

with about 600 ml for the gorilla, 450 ml for the chimpanzee, and 400 ml for the orangutan. We must remember, though, that these animals do not all have the same body size, and when allowance is made for relative size, we find that the chimpanzee has the second-largest brain, followed by the orangutan, and then the gorilla. Man is way out in front of them all. Relative brain size is assessed using a quantity called the encephalization quotient, written EQ (*not* to be confused with IQ). We need not be concerned here with how the EQ is calculated (an explanation appears in the reference notes for Chapter 14); suffice it to say that it gives a comparison between the brain size of the animal under consideration and the average brain size for that particular class of animal. For example, our EQ is about 7.5, which means that our brain is about 7.5 times larger than the average brain size for mammals of similar body weight. The EQ of the chimpanzee is about 2.3, that of the orangutan is 1.8, while that of the gorilla is 1.6 (data from Jerison, 1973). We will be referring to EQ again when we discuss our fossil relatives.

The primates, then, bear the stamp of having evolved in the trees, and our primary concern in this chapter will be to trace the evolutionary stages in man's descent to the ground. Before we begin looking at this fascinating story, we need to have some idea of how our skeletal anatomy compares with that of other primates. Our closest living relatives are the apes, and we will therefore make a comparison between man and apes. In doing this we are not implying that we evolved from apes; instead, we believe that man and apes had a common ancestor.

Modern apes, which belong to a group called the pongids, have diverged from the group to which we belong, called the hominids, but we believe that our more remote ancestors were apelike rather than monkey-like or lemur-like, simply because of the closer similarity between man and apes. The skeletal anatomy of the apes, then, gives us a good idea of the ancestral condition. To give an analogy, the automobile engine has more in common with the steam engine (pistons, piston rods, cylinders, reciprocating action, induction valves, to name a few features) than it does with a jet engine or a rocket engine, and we get a better idea of the ancestral car engine by looking at a steam engine than we do by looking at either one of the other two engines.

When we make our comparison between man and the apes, we should bear two features in mind which set man off from the other primates, and, for that matter, from all other animals:

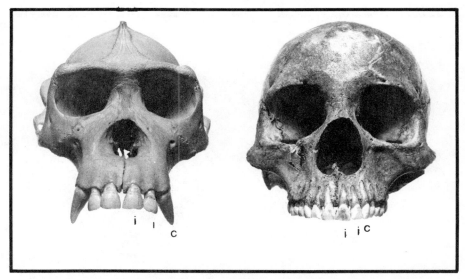

Front view of the skulls of a chimpanzee (left) and man. Notice the larger incisors (i) and sharp canines (c) in the chimpanzee.

we walk on our hind legs with our vertebral column held upright, and we have an enormous brain (both absolutely, and relative to our body weight). Both of these features can be recognized in the skeleton. Because most fossil hominids are represented by teeth, jaw fragments, and the occasional skull, we will pay particular attention to these structures in making our comparisons. Unfortunately the anatomy of teeth and jaws, and even of skulls, is rather complex stuff, but we will try to make the explanation as painless as possible.

A few years ago the Brooke-Bond Tea Company in England ran a series of TV commercials featuring a pair of chimpanzees, suitably attired, of course, and enjoying their tea like good Englishmen. At some point during their performance they would part their lips in a broad simian grin, revealing the most enormous front teeth and vicious-looking canines. Apes have very large incisors, especially the first incisors in the upper jaw, which exceed even the cheek teeth in size. The fact that the upper incisors project forwards makes them look even more prominent. The lower incisors do not project forward, but, being so large, they project well above the level of the biting surfaces of the cheek teeth. Likewise with the canine teeth, which are fairly sharply pointed. Our incisors and canines, in contrast, are relatively small, and neither project forwards, nor

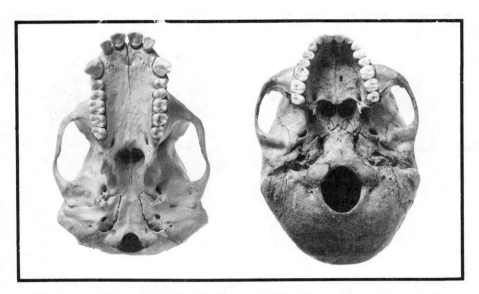

Comparison of the skulls of a chimpanzee (left) and man, from below. Notice the gap (diastema) between the second incisor and the canine teeth in the chimpanzee. Notice also that in the chimpanzee the cheek teeth lie in parallel rows, whereas in man they lie in a curve.

above the level of the cheek teeth. Furthermore, our canines are chisel-shaped rather than pointed, and look much like our incisors. You can check all of these points for yourself by running your tongue across your teeth. Male apes have relatively larger canines than females, so that it is possible to determine the sex of an ape just by looking at its teeth, but there is no similar sexual difference in our own species.

Most of us get our wisdom teeth, the third molars, when we are in our late teens or early twenties, by which time we have long since got all of our other permanent teeth. Apes, however, are different in that their last permanent teeth are their canines. One way of determining whether a human skull belonged to an adult is to check whether the wisdom teeth have erupted, but for apes one would check the canines. If we took a close look at a smiling chimpanzee, or any other ape for that matter, we would see a small gap, or diastema, in the upper jaw between the canine teeth and the incisors. We do not have a diastema. The last dental comparison we need to make concerns the slope of the upper and lower tooth rows. In the apes the cheek teeth and the canine tooth of the left and right sides lie approximately in a straight line, and these are more or less

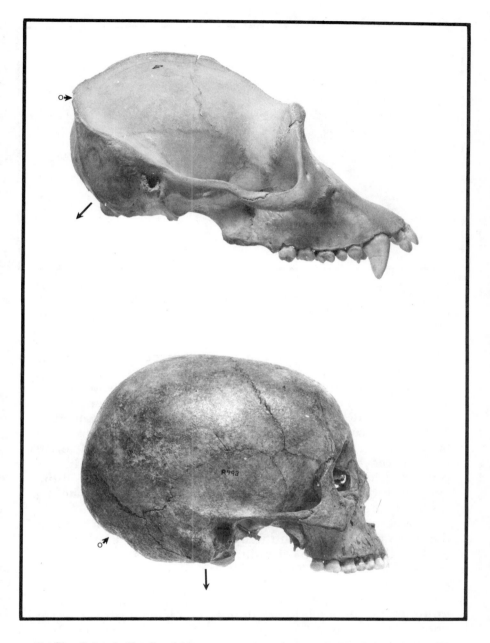

Profile of the skulls of a chimpanzee (top) and man. Notice that the occipital protuberance (o) is much higher in the chimpanzee than in man. Also notice that the foramen magnum points directly downwards (arrow) in man but obliquely backwards in the chimpanzee.

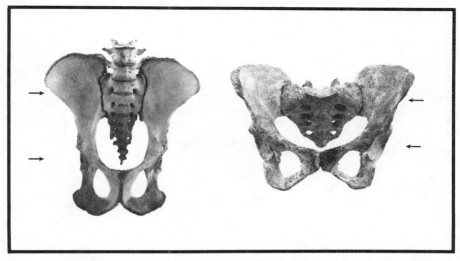

Comparison of the pelvis of a chimpanzee (left) with that of man. Notice the narrow hips in the chimpanzee and also the greater distance between the level of attachment of the vertebral column (top arrow) and the hip joint (bottom arrow).

parallel to one another. The distance between the canines is therefore about the same as the distance between the last molars. This contrasts sharply with our own teeth, which lie along a gently curving arch, and the distance between our canines is considerably less than the distance between our last molars. Before leaving the chewing apparatus, we must just take a look at the lower jaw. If we looked on the inside of an ape's lower jaw, just at the point where left and right halves come together, we would see a small shelf, which is about big enough to balance a pea on. This is called the simian shelf, a feature that man lacks.

Viewed in profile, the cranium of an ape comes to a peak, at the back, on a level with the upper margin of the orbit. This peak, called the occipital protuberance, marks the apex of a crescent-shaped area. This is called the nuchal (Latin for neck) area, and is the point of attachment of those neck muscles which are used primarily for supporting the head. In man, the occipital protuberance is much lower down, on a level with the lower border of the orbit. Consequently, the nuchal area, which is less extensive than in the apes, faces downwards rather than downwards and backwards. This different arrangement is correlated with our upright posture; not only do we need less extensive muscles to support our heads, but these lie in an essentially vertical rather than an oblique direction. The position of the foramen magnum — the hole through which the spinal cord passes to the brain — also differs between apes and man. Like the nuchal area, the foramen faces vertically downwards in man, but is inclined obliquely downwards and backwards in apes.

Although apes are able to walk in an upright position, they are not very good at it, and they cannot keep it up for long before reverting to their semi-upright position, propping themselves with their front legs. The reason apes have trouble at walking upright is simply that they are not built for it. Their hips are fairly narrow, because the ilium (upper bone) is longer than it is broad and is attached to the backbone well above the level of the hip joint. This makes it rather difficult for them to balance with their back held vertical. Our hips, in contrast, are wide, the ilium being much broader than it is long. Most of the increase in width is due to an increase in the posterior region of the ilium, which provides a large attachment area for the buttock muscles, which are used in swinging the leg back. Furthermore, the hips are splayed out towards the top, so that the pelvis forms a bowl to help support the internal organs. The flaring out of the hips is especially prominent in women, and is correlated with child-bearing. The point of attachment of the pelvis to the backbone is low down, essentially on a level with the hip joint, and this increases the stability of the

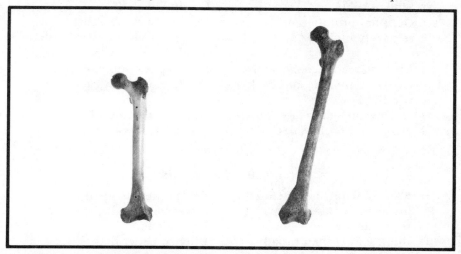

Front view of the left femur of a chimpanzee (left) and man. The shaft of the femur slopes outwards in man because of the inclination of the knee joint to the shaft.

A Comparison of Skull and Dental Characteristics and of Some Postcranial Characteristics in Apes and Man

Apes	*Man*
Skull and dental characteristics	

1. Incisors large.	Incisors small.
2. Upper incisors project forwards.	Upper incisors do not project fowards.
3. Canines large.	Canines small.
4. Canines project well beyond level of other teeth.	Canines on same level as other teeth.
5. Canines conical, tapering to a sharp point.	Canines chisel-shaped rather than pointed.
6. Canines much larger in males than in females.	Canines same size in males and females.
7. Canines are last permanent teeth to erupt.	Third molars are last permanent teeth to erupt.
8. Diastema (gap) in upper jaw between canine and incisors.	No diastema in upper jaw between canine and incisors.
9. Canine and cheek teeth lie in a straight row; rows parallel with one another.	Canine and cheek teeth lie on a gently curving arch.
10. Distance between canines similar to distance between last molars.	Distance between canines much less than distance between last molars.
11. Simian shelf.	No simian shelf.
12. Occipital protuberance relatively high.	Occipital protuberance low, on a level with lower border of orbit.
13. Nuchal area extensive and faces downwards and backwards.	Nuchal area not extensive and faces downwards.
14. Foramen magnum faces downwards and backwards.	Foramen magnum faces vertically downwards.

Postcranial characteristics

15. Shaft of femur not inclined inwards.	Shaft of femur inclined inwards.
16. Hips narrow.	Hips broad.
17. Ilium longer than broad.	Ilium broader than long.
18. Hips attach to backbone well above level of hip joint.	Hips attach to backbone close to level of hip joint.

trunk on the legs. Another feature that helps us in our balancing act is the way we place our feet, well in towards the center line of our body. The footprints we leave in the snow almost fall along a straight line, and if we measured the width of our tracks, from the center of one footprint to the center of the other, we should find that it was less than the width from one hip joint to the other. How is this achieved? If we look at the distal (lower) end of the human femur (thigh bone), we can see that the paired condyles, which articulate with the tibia (shinbone), are set at an angle to the shaft of the bone. Therefore, if we held a femur upright and placed the two condyles flat on a table, the shaft would be inclined outwards. Therefore, when the two femora are popped into their respective sockets in the pelvis, the distal ends are inclined towards each other, and this is why our knees brush past each other when we walk.

Having finished our comparisons between apes and man, we are ready to take a look at the fossil evidence of our ancestry.

The chain of events leading up to some of the fortunate accidents of fossil discovery must rank as highly improbable when the odds are assessed. Consider the circumstances that led to the discovery of the first "man-ape" specimen, *Australopithecus africanus*. Towards the end of 1924, Miss Josephine Salmons, a student of anatomy at the University of Witwatersrand in South Africa, brought to her professor, Raymond Dart, a fossil primate skull which she had obtained on loan from a mining company. The fossil in question had been blasted out of a limestone formation at a place called Taungs, and Dr. Dart, though not a paleontologist himself, was sufficiently interested by the specimen to consult a colleague in the Department of Geology. As it happened, the geology professor was just about to visit the Taungs area himself, to investigate the geology of the lime deposits. He agreed to examine the site where the skull had been found.

When the geologist got back from his trip, he visited Dr. Dart and gave him an assortment of fossils which he had obtained from the limestone formation. Among these was the natural cast of the brain of a primate. Natural brain casts are rare — they are formed when sediments fill a cranium and become consolidated into rock — and they are especially rare for primates. We can imagine Dr. Dart's elation when he discovered that the brain cast matched up with the remains of a skull which was still largely embedded in limestone. When he had carefully removed the skull from its stony shroud he found

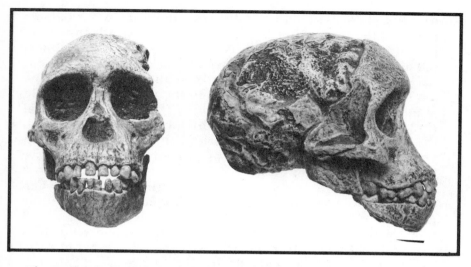

The Taungs skull, the first discovered australopithecine. This is actually a juvenile individual of the gracile species, Australopithecus australis.

that he had not just an ape fossil, but a fossil that possessed many hominid features. As in man, the cheek teeth and the canine tooth in each half of the jaw lay along a gently curving arch, rather than being arranged in straight rows as in the apes. The incisors and the canines were small, and the fact that the permanent canines were present showed that they had erupted before the wisdom teeth, which had not yet erupted. The skull obviously belonged to a juvenile. Not only were the front teeth small but the upper incisors did not project forwards, nor did the canines project beyond the level of the cheek teeth. In all of these dental features, the fossil corresponded to the human condition. (There was a small diastema between the upper canines and the adjacent incisors, but Dart noted that this was smaller than the gap between the incisors, and was due to immaturity — our own children have obvious gaps between their milk incisors.) Dart looked at the lower jaw and found there was no simian shelf — another hominid feature — but even more interesting was the fact that the nuchal area appeared to face more directly downwards than it did in the apes. The implication of this was that the animal may have walked erect. Another particularly interesting feature was the large size of the brain. Dart compared this with the brain of an adult chimpanzee, concluding that it was, if anything, slightly larger. When account is taken of the fact that Dart's specimen was a

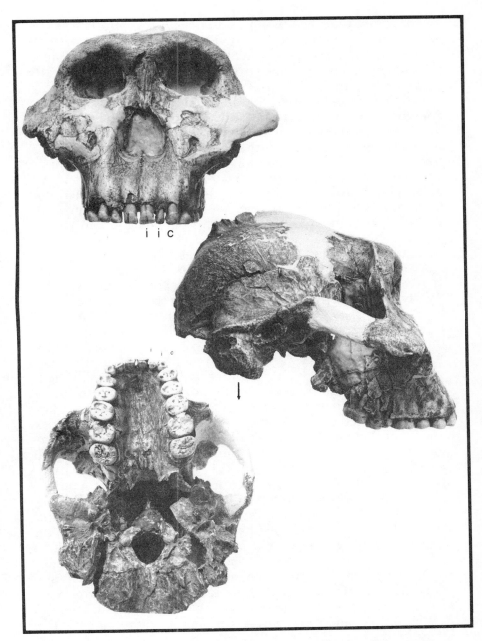

The skull of an adult individual of Australopithecus robustus. *Notice the following man-like features: relatively small incisors; canines small and not pointed; no diastema; teeth lie in a curve; the foramen magnum points directly downwards.*

juvenile, and that the human brain does not attain its full size until adulthood, this fact is remarkable. Here was a fossil primate which seemingly walked erect and which had a relatively large brain. Dart wrote that "It is therefore logically regarded as a man-like ape."

Had Dart found a fossil that bridged the gap between some apelike ancestor and man? This question has been widely discussed since the announcement of the discovery, and while some scientists agreed that the specimen did possess hominid features, others dismissed it as being nothing more than an ape. One of the problems was that Dart's specimen was a juvenile — it still had milk teeth — and it was suggested that its hominid features were due to its immaturity. If a juvenile ape skull is compared with that of a human, the similarities are striking.

A large number of australopithecines have been discovered since Dart's time, and we now have a good knowledge of their anatomy. Two basic types are recognized: gracile (slender, lightly built) and robust. The Taungs skull belongs to the gracile type, *Australopithecus africanus*, while almost all of the robust types are referred to the species *Australopithecus robustus*. One of the most thorough studies of their anatomy was made by the British anatomist W. E. Le Gros Clark, and although this work was done back in 1950, his findings and conclusions have only been strengthened by later studies. Working only with skulls, and paying particular attention to their teeth, Le Gros Clark made a detailed comparison between australopithecines, apes, and man. Fourteen of the skull and dental features he compared were included in the table that appears earlier in this chapter, and these cover most of the points considered by Le Gros Clark (he also included features of the milk teeth, but we will not). The australopithecines correspond to man in all fourteen features. Furthermore, they also correspond to man in the milk-teeth features assessed by Le Gros Clark, and also in the structure of the hip and the femur. The australopithecine ilium is broader than it is long, its point of attachment with the vertebral column is low down, and the distal end of the femur is angled to the shaft, giving the knock-kneed posture to the legs. There is therefore no question that the australopithecines are hominids, and the consensus among paleontologists is that they walked erect.

One of the most obvious differences between *Homo* and *Australopithecus* is the relatively enormous size of the cheek teeth in the latter. Even in the Taungs skull, which is only a youngster, the cheek teeth appear disproportionately large, and this is especially so in the robust species of the genus. Little wonder that one of the robust specimens collected by the legendary Louis Leakey should have been nicknamed "Nutcracker Man" (*Australopithecus boisei*). There could be no mistaking the robust australopithecines (*Australopithecus robustus* and *Australopithecus boisei*) for modern man; not only are the teeth massive compared with our own, but the skull is exceedingly heavily built. The brow ridges are great buttresses running across the top of the orbits, and there is a ridge of bone running along the top of the skull, called the sagittal crest, which is also seen in the male gorilla. The zygomatic arches, the cheekbones to which the jaw muscles attach, are great arches of bone which bow out at the sides of the skull like a pair of handles. The gracile australopithecines (*Australopithecus africanus*) have less massive skulls with relatively smaller teeth, less well-pronounced brow ridges, no sagittal crest, and much more slender zygomatic arches.

The hallmark of *Homo sapiens*, the wise man, is his large brain, which, we have seen, ranges from about 900 ml to 2200 ml. Australopithecines had much smaller brains; estimates range from 430 ml to 530 ml, which is comparable to the brains of apes. However, we must remember that brain size, taken in isolation from body size, has little meaning, and when we consider the EQ values for australopithecines, we find that they are high, ranging from 3.2 in *A. robustus* to 4.3 in the gracile *A. africanus*. Apes, by contrast, have EQ values ranging between about 1.6 and 2.3. Quite clearly the australopithecines had relatively much larger brains than apes, approximately twice as large, but they had not reached the level of brain development seen in man. When we come to discuss the earlier hominids, and then the later ones, we will see that there was a gentle trend towards increasing brain size.

What time period were the australopithecines walking the earth? The gracile form, *A. africanus*, appeared first, about 2.8 million years ago, becoming extinct about 2 million years ago. The robust forms, *A. robustus* and *A. boisei*, which probably evolved from the gracile form, appeared about 2.1 million years ago, and survived until less than a million years ago.

Some scientists have claimed that the robust and gracile forms are really just male and female members of the same species, arguing that a similar degree of sexual dimorphism can be seen today in the gorilla. The fact that there is a geological time difference between the two, however, rules this possibility out, and the sexual dimorphism argument has been widely rejected.

Here we are, more than halfway through the chapter, and we have not asked how the creationists interpret these data. Dr. Gish is particularly interested in hominid fossils. He tells us that the australopithecines had small brains, averaging 500 ml or less, which is in the range of the gorilla. He tells us that they "had ape-like skulls and jaws." Citing papers by Richard Leakey, he suggests that they were long-armed and short-legged and that they propped themselves up with their front limbs like apes when they knuckle-walked instead of walking erect. Dr. Gish also tells us that the robust and gracile forms are males and females of the same species. Finally, he casts doubts upon the validity of certain dental features that have been used to assign the australopithecines to the hominid family. Citing the works of C. J. Jolly and D. R. Pilbeam, he points out that there are certain baboons, called gelada baboons (genus *Theropithecus*), which have small incisors and canines and a short, deep face like australopithecines. Since this baboon is clearly not a hominid, he quite reasonably argues that these dental features cannot be used to characterize hominids.

We will now deal with each of these five points in turn: small brain size, apelike skulls and jaws, non-erect walking, sexual differences, and doubts regarding dental features.

We have already shown that although the australopithecines had small brains compared with ours, their relative brain size, as measured by the EQ value, was considerably larger than that of apes. We have also seen that the skulls and jaws of australopithecines, especially the gracile ones, are manlike rather than apelike.

While it is true that Richard Leakey suggested that "The Rudolph australopithecines, in fact, may have been close to the 'knuckle-walker' condition, not unlike extant African apes" in the paper cited by Dr. Gish, it is true to say that his is a minority view. This tends to be confirmed by Leakey's changed position in his book *Origins*, published six years after this article, where he says that "these creatures [australopithecines] were quite definitely habitual upright walkers and there is no evidence to support the suggestion that they would occasionally pound along on all fours in the knuckle-walking style of chimpanzees or gorillas."

Whether or not the robust and gracile species of *Australopithecus* represent males and females of the same species has little bearing on our discussion of australopithecine relationships, but we should point out that while Richard Leakey did cast doubts on the two-species hypothesis in the paper (1971) cited by Dr. Gish, in a later paper (1976) he noted that "I am of the opinion that evidence for two species of this genus [*Australopithecus*] can be established with some conviction in the Koobi Fora Formation." (The Koobi Fora is the name of a fossil locality on Lake Turkana in Kenya.) Even more significant than this change of view is the fact that a time difference separates the gracile and robust forms, as noted above.

Gish's references to the gelada baboons sounds a little disturbing, because it seemingly destroys the validity of the dental characteristics which we have used to distinguish the hominids from the other primates. However, when we look at a gelada baboon we see that only two dental features parallel the hominid condition: the incisors are relatively small compared with the cheek teeth, and they tend to be vertical rather than to point forwards. In all the other dental features the gelada baboon compares with non-hominids: the cheek teeth and canines on the left and right sides lie in a straight row, and these rows lie essentially parallel to one another; the distances between the canines and the last molars are approximately the same; the canines are long and project beyond the level of the cheek teeth, especially in males; the canines are pointed; and there is a diastema between the canine and the incisor teeth. The suggestion that the gelada baboon possesses dental features that undermine the hominid status for australopithecines is therefore unfounded.

Our discussion of the australopithecines has established the following points about them:

- Australopithecines had relatively large brains, much larger relative to their body size than those of apes, and in this regard they approach our own condition.
- Australopithecines have the same cranial, femoral, and pelvic features as we do, features that are associated with erect posture.
- Australopithecines share a large number of skeletal features with us.
- Australopithecines differ from us in the following features: smaller brain, large molar teeth, bony brow ridges, sagittal crest (in the robust species).
- The australopithecines so far discussed appeared about 2.8 million years ago.

The australopithecines, then, appear to be the ancestral group from which our own species evolved. We will now look at the next steps in hominid evolution to see whether we can establish a sequence of evolutionary change leading to *Homo sapiens*.

In 1964 Louis Leakey, in collaboration with P. V. Tobias and J. R. Napier, announced the discovery of a new species of man from the famous Olduvai Gorge fossil locality in Tanzania. The new find was named *Homo habilis*, the handy man, because he was credited with having made the stone tools that are found in the early part of the Olduvai sequence. Louis Leakey had formerly believed that the stone tools of the Oldowan culture had been made by *Australopithecus boisei* ("nutcracker man"), because at the time of the discovery of this robust species of australopithecine (in 1959 by Louis's wife, Mary), it was the only hominid known at that geological level.

The description of this new species of *Homo*, the first discovered since *Homo erectus* (formerly known as *Pithecanthropus erectus*) in 1896, was reason alone to make the scientific world sit up and take notice, but this new material was dated at 1.7 million years, pushing man's evolutionary history back more than twice as far as it was before. It was important that such a revolutionary new find should be firmly established as a valid species; therefore, Leakey and his associates had to be sure that the specimen really did belong to the genus *Homo*. Up till that time the distinguishing feature of *Homo* had been brain size, and the cut-off point had been variously set at between 700 ml and 800 ml. Any fossil with a brain size equal to or exceeding this value was classified as *Homo*. The dangers of defining a group on one characteristic alone are well known, and Leakey and his co-workers therefore proposed additional diagnostic features to define the genus *Homo*, including:

- Forelimb shorter than hind limb
- Thumb well developed and fully opposable (that is, the thumb is able to touch all of the other fingers)
- Cranial capacity generally greater than that of australopithecines, varying from about 600 ml to 1600 ml
- Facial region never concave, as it is in australopithecines
- Molar teeth generally small compared with those of the australopithecines

The *Homo habilis* material comprised five specimens, largely fragmentary but including an incomplete cranium with associated teeth, associated hand and foot bones, and two mandibles. Estimates of brain size made by P. V. Tobias that same year gave values of between about 640 ml and 720 ml. This is intermediate between *Australopithecus* and *Homo erectus*, but the cranium that Tobias had to work with was so fragmentary that these results are really rather tentative. Indeed, the scientific community as a whole was rather skeptical about *Homo habilis*, many preferring to interpret the material as being australopithecine. For Leakey, though, the evidence of the cranial capacity was enough to convince him that his new discovery really was the earliest man.

As a youngster, Louis Leakey's son Richard was not very interested in fossils, nor was he interested in going to university and becoming a paleontologist like his father. Instead, he became a safari guide for visitors to East Africa. As a result he gained a good knowledge of the bush and of organizing field parties, and so it was, in 1967, that he took responsibility for a fossil-collecting expedition to Ethiopia, to a place called Omo, close to the border with Kenya. From that time on he gave up his safari business and became a professional paleontologist. After the Omo expedition Richard Leakey moved south to Koobi Fora in northern Kenya, set up a permanent camp, and began a collecting program with other professionals. His group made a number of interesting finds over the next few years, and Richard Leakey's reputation as a hominid specialist grew. Most of the hominids they found were robust australopithecines; then, in 1972 Richard Leakey announced the discovery of a very interesting and almost complete skull, designated number 1470.

Was 1470, provisionally dated at 2.9 million years old (but, as we shall see later, the revised dating is less than 2 million years), another australopithecine or was it a man? The high-crowned and well-rounded skull, which was remarkably thin-walled and lacking in large, bony buttresses, did not look as if it belonged to *Australopithecus*. Even more impressive was its large cranial capacity, estimated at 775 ml. This was no australopithecine, and Richard Leakey had no hesitation in pronouncing that he had found the oldest known specimen of *Homo*. He was reluctant to attach a specific name to it, though, and only later did he decide that his new specimen should be referred to as *Homo habilis*. His father's belief in the validity of *Homo habilis*, and in the great antiquity of man, was vindicated. Louis Leakey died that same year, happy in the knowledge that *Homo habilis* was firmly established as a valid species.

What have the creationists to say about *Homo habilis*? First we should point out that the first edition of Dr. Gish's book was published the same year as Richard Leakey's discovery, the second edition appearing just one year later. Dr. Gish was therefore unable to read the various papers that have been published regarding *Homo habilis*, but the fact that he does have a brief discussion of the implications of this discovery in his book is much to his credit. Both Drs. Gish and Morris leap upon the apparent anomaly in timing: the earliest man, *Homo habilis*, seemingly appeared *before* the australopithecines. How, they reasonably ask, can a descendant be older than its assumed ancestor? To resolve this apparent dilemma we have to take a trip to Ethiopia and meet a paleontologist named Donald Johanson.

Donald Johanson was still a graduate student in 1970 when he went on his first trip to Africa. After meeting up with his doctoral supervisor, Clark Howell, in Nairobi, the two spent a few days together before flying on to Omo to spend three weeks looking for fossils with Howell's field crew. It was while they were in Nairobi that Johanson met Louis Leakey's wife, Mary, and renewed his acquaintance with Richard Leakey, who was visiting from his camp at Koobi Fora. He showed Johanson some of the latest australopithecine finds he had made; Johanson was suitably impressed.

Johanson spent the next two summers collecting at Omo, received his Ph.D., then obtained a grant to set up his own collecting operation in northern Ethiopia. Johanson's camp was at a place called Hadar in the area known as the Afar Triangle. A previous reconnaissance trip to Hadar had convinced Johanson that this was a rich fossil locality, and that the advantage it had over Omo, to the south, was that it was geologically older, dating back more than three million years. Also, because the sediments are frequently interbedded with thin layers of volcanic ash, ideal for radiometric dating, it would be possible to get good absolute dates for the various fossil beds. So confident was Johanson of finding some really interesting hominid fossils there that he had a bet with his friend Richard Leakey.

The first Hadar field season got under way in the fall of 1973, and Johanson found his first important fossil: a knee joint. Now a knee joint does not sound especially interesting, but this one was, for two reasons. First, the condyles at the distal end of the femur were inclined to the shaft, knock-kneed fashion, showing that the joint belonged to an erect walker — it was hominid. Secondly, the fossil was some three million years old — the oldest hominid — older even than Richard Leakey's *Homo habilis.* The bones were fairly small, much smaller than in a human knee joint; the question was, to what little hominid did they belong? If they belonged to an australopithecine, it was the smallest one yet found, probably about a meter tall.

The following year Johanson and his crew found three hominid jaws, the oldest material of its kind ever found. Man or australopithecine? The jaws possessed features common to both. Johanson invited Richard Leakey and his mother to visit his camp and give him an opinion on the new finds. When they saw them they were both agreed: the jaws looked as if they belonged to *Homo*. They then discussed the other animals that had been found at the Leakeys' site at Koobi Fora and at the Hadar site. Although the layer from which Richard Leakey's famous *Homo habilis* skull had been collected was supposedly of about the same age as the Hadar locality from which Johanson's three jaws had come, the associated animals were different. Leakey's crew were apparently already a little uncertain about the 2.9-million-year age assigned to *Homo habilis,* and when Johanson was asked for his opinion, he suggested that the age was only about 2 million. Subsequent radiometric dating of uncontaminated samples of volcanic ash from the layer just below that from which the skull had been taken gave a date of 1.8 million years, an age that has now been generally accepted. The Leakeys left the following morning, and on the very next day Johanson made the discovery of a lifetime. He found Lucy.

Lucy, named after the Beatles song which was popular in the Johanson camp that season, was an almost complete skeleton — the most complete Pliocene hominid ever found. What was she? Lucy had the legs and hips of an erect walker, of that there was no doubt. Unfortunately, although her jaws were well preserved, the skull was incomplete and it was not possible to estimate the cranial volume, but the implications were that it was small, *Australopithecus*-sized rather than man-sized. Did Lucy belong to the same species as the knee joint of the previous season, and the three jaws found days before? Only time could tell.

More hominid material was found at Hadar during the third and fourth field season. Meanwhile, Mary Leakey, working at a place called Laetoli in Tanzania, some 1600 km to the south, was finding hominid material similar to that found at Hadar. The age of this material was about 3.7 million years. Working with Mary Leakey was a paleontology graduate student named Tim White, who was later to collaborate with Johanson in the study of the Hadar material. White and Johanson met, in Nairobi, at the end of Johanson's third field season at Hadar, during a visit to the Leakeys. After some discussion of the material from the two sites, White made a statement that startled Johanson: "I think your fossils from Hadar and Mary's fossils from Laetoli may be the same." White showed some of them to Johanson — they appeared almost identical.

While Johanson was working his fourth season at Hadar, a monumental discovery was made at Laetoli: animal tracks, imprinted into what had been a fresh fall of volcanic ash some 3.7 million years before. The significance of the discovery was not recognized until the following year, 1977, when further excavation revealed a number of tracks that looked as if they had been made by humans. The footprints clearly showed a heel imprint, set off from a ball by an arch. There was an imprint of a big toe, pointing directly forwards, not out to the side as it would have done if the print had been made by an ape. Here was evidence that there were erect, bipedal walkers living in Africa 3.7 million years ago. Now the pieces of the puzzle began to fit together.

Johanson and White worked on their analysis of the Hadar and Laetoli specimens through the summer of 1977 and into 1978, publishing their findings in papers that appeared in 1978 and 1979. Mary Leakey was to be one of the authors of the descriptive paper that was published in 1978, but she was apparently so angered by the results of the analysis, announced by Johanson at a scientific meeting in Sweden in 1978, that she withdrew her name.

What was it about Johanson and White's results that caused the rift with the Leakeys? In the first place, they concluded that Lucy and the other Hadar material, together with the hominids from Laetoli that looked almost identical to the Hadar specimens, all belonged to the same species, for which they proposed the new specific name *afarensis*. The official description of the new species was therefore partly based upon the Laetoli material, and this may have caused some of the problem, encroaching, as it were, on the Leakeys' territory. Perhaps more significant still was Johanson and White's conclusion that their new species was an australopithecine. *Australopithecus afarensis* was therefore the oldest hominid, and they suggested that it was the ancestral stock from which both man (*Homo habilis*) and the other australopithecines evolved. The Leakeys, however, firmly believed that the genus *Homo* was the oldest hominid. If Johanson and White had concluded that their new species belonged to the genus *Homo* rather than to *Australopithecus*, the rift might never have developed. Actually Johanson had originally believed that his Hadar material did belong to *Homo*, and the fact that it was difficult to decide between the two genera goes to show how closely one merges into the other. As we have seen before, it is almost impossible to know just where to draw the line when dealing with a transitional series.

So far we have seen that the earliest hominids were erect bipeds that lived more than 3 million years ago. (Radiometric dating at Laetoli gave ages of about 3.7 million years, and sixteen separate tests on uncontaminated ash samples from Hadar gave similar dates for the Hadar material.) These hominids are classified as australopithecines and estimates of their cranial capacity range from about 380 ml to 450 ml. This overlaps with that of apes, but, as we have said before, it is *relative* brain size that is important, and australopithecines had brains that, relative to their body sizes, were considerably larger than those of apes. Australopithecine brains, however, are still small compared with those of man. All of this is contrary to the earlier ideas of hominid evolution which held that large brains came before erect bipedal posture. It is interesting to note that while Lucy's pelvic girdle is closely similar to that of man, the birth canal is considerably smaller. The reason for a large birth canal in man is to accommodate the relatively large head of the infant, but australopithecine babies did not have such large heads. While it might seem like a substantial jump from *Australopithecus afarensis*, with a cranial capacity of about 400 ml, to *Homo habilis* at about 700 ml, we have to remember that the former was much smaller than the latter. Aside from differences in cranial capacity and body size, the differences between the two are minor.

What about the stages leading from *Homo habilis* to *Homo sapiens*? When the Dutchman Eugene Dubois found a thick skull cap, which had heavy brow ridges and a low crown, in a river bank at Trinil in Java towards the close of the nineteenth century, he suspected that he had found the elusive "missing link" between apes and man. The femur, which was found near by, and which he considered to be part of the same individual as the partial skull, was very human-like, and obviously belonged to a creature that walked erect, but what sort of a creature was it? Dubois estimated the cranial capacity to be about 900 ml, considerably larger than that of any ape but rather small for a man. This, together with the heavy build of the skull, and the large size of the few teeth that he had found close to the skull, convinced him that this was indeed an intermediate fossil linking apes and man. He called it *Pithecanthropus erectus* (Greek *pithekos*, meaning ape; *anthropos*, meaning man).

Dubois was disappointed that so few scientists agreed with him: many concluded that the skull belonged to an ape, a few believed it to be human, and fewer still agreed that it was intermediate between apes and man. He contrasted this un-acceptance of his discovery with the discovery of Neanderthal man, whose affinities had never been questioned. (Neanderthal man is recognized as a primitive member of our own species, *Homo sapiens.*) As we saw at the beginning of the chapter, Neanderthal man also had a thick, low-crowned skull, with a heavy brow ridge, but his cranial capacity was somewhat larger, something over 1000 ml.

Many more specimens like Dubois's have been found since the turn of the century, and these are now all referred to as the species *Homo erectus*. *Homo erectus* has been found in many parts of the world — Europe, Africa, China, and the Middle East — and includes specimens that have been referred to as Java man (the original specimen found by Dubois), Peking man, and Heidelberg man. We now know that the cranial capacity of the species overlaps broadly with the lower end of our own range, and, aside from the heavier build, especially of the skull, there are no significant differences between it and our own species. There is, therefore, a gentle progression from *Homo erectus*, which lived from about 1.5 million till about 200,000 years ago, to *Homo sapiens.* Many more Neanderthal specimens have also been found, mostly in Europe, but also in Africa, the Middle East, and China. It is interesting to note that the first Neanderthal specimen and the first specimen of *Homo erectus* were both dis-eased, various parts of their skeletons having bony outgrowths such as occur in arthritic conditions.

And what do the creationists have to say about *Homo erectus* and Neanderthal man? Dr. Gish goes to some lengths in his discussion of *Homo erectus* to establish that there were suspicious circumstances surrounding the discovery of both Java man and Peking man, and the notion of fraudulence is reinforced by his telling the story of the Piltdown hoax. Dr. Morris also touches upon the circumstances of the discovery of Peking man, but he goes on to make the valid point that other specimens of *Homo erectus* have been found at various locations around the world. He concludes that although *Homo erectus* may well have been a true man, it was somewhat degenerate in size, "possibly because of inbreeding, poor diet and a hostile environment." Neither Dr. Gish nor Dr. Morris have very much to say about Neanderthal man, aside from the fact that the original material was arthritic and that Neanderthals are like modern man. Dr. Morris ends his discussion on the subject by saying that as far as the fossil record is concerned, man has always been man, and that "There are no intermediate or transitional forms leading up to man, any more than there were transitional forms between any of the other basic kinds of animals in the fossil record." Our survey of the fossil evidence of hominid evolution has shown that this is just not true, any more than it is true for the other transitional fossils that we have discussed.

The Emperor's New Clothes

 WHEN THE EMPEROR in Hans Christian Andersen's tale put on his brand-new "robes," none of his courtiers wanted to show their ignorance by pointing out that he was actually wearing nothing at all. Instead, they complimented the Emperor on the magnificent cut and the dazzling colors of the nonexistent garment, and all agreed that he had made a most splendid and wise purchase. I was reminded of that story when I sat and listened to the orations of a creationist at a public lecture.

Take a bit of science, like thermodynamics, that most people know nothing about, add some biochemistry and a dash of mathematics, season with a few jokes and a pinch of southern charm, and serve with the righteous fervor of an evangelist, and you have your audience eating out of your hand. On this occasion several people pointed out that the Emperor had no clothes on, but many more were dazzled by the wondrous robes.

To appreciate why an argument is riddled with holes, especially an argument that is eloquently presented, requires a sound knowledge of the subject, and here is the main reason the creationists have made such gains in the last few years. We may be living in an age of science but the majority of people, the huge majority, know very little about the subject, and our educational institutions have to bear much of the responsibility for this. I fear that we do not always do the very best job of teaching science in our schools, or, unfortunately, in our universities. This is especially true for the subject of evolution, which is usually either treated superficially, or not treated at all. My thirteen- and sixteen-year-old daughters, both in the public school system, have had no exposure to the subject as far as I can ascertain, and there are precious few university courses that make a good job of presenting the evidence for evolution. It therefore always amuses me when I hear the creationists demanding equal time for the teaching of their views.

The equal-time plea has actually worked wonders for the creationists, appealing as it does to people's sense of fair play. What could be more reasonable than presenting both views on the controversial question of how living organisms came into being? This sentiment is made all the more appealing by the creationists when they point to all the arguments and disagreements presently going on among evolutionists. People who are unfamiliar with the subject and the issues would readily be swayed by this line of reasoning, but not those with an understanding of science. Whatever label the creationists want to pin on their religious beliefs, it is most certainly *not* science, and has no place in the science classroom. I hope this point has been adequately demonstrated in this book. Furthermore, the current arguments going on among evolutionists are not over the question of whether evolution has occurred — few scientists doubt this — but over the question of the mechanisms of evolution.

Apathy is one of the sins of which we are all guilty in this day and age. We lead such busy lives and have little time to stand and stare. The creation-evolution controversy must rank fairly low in our scale of priorities, well below our concerns for the economy, and further still below our concerns over the nuclear threat. It would be all too easy to shrug it off as an issue of marginal importance, an issue that will probably all blow over in any case. But I would like to suggest that it is not such a small issue, and that it really is not going to go away. I am not unduly troubled that my two daughters are not being taught about evolution in their science classrooms, but I am concerned that they might be exposed to a hodgepodge of ignorance and half-truths presented to them as a "scientific" alternative to evolution. As I see it, the evolution issue is just the thin edge of the wedge. If creationism is successfully legislated into our school system, what will come next? Why should the fundamentalists stop short at biology? Why should they not also have the right to teach their beliefs in a 10,000-year-old universe and in light rays that distort to fit the "facts." Should the creationists have their way, I believe we would see science revert back to the mystical art form that it was during the dark ages.

And science will not be the only casualty, for science is only one branch of learning, and the essence of all learning — be it science, the arts, or the humanities — is to expand the frontiers of human knowledge. With their literal, fundamentalist views, the majority of creationists do not place a high value on the freedom of inquiry and expression necessary to the growth of learning. To dismiss them as harmless, simple-minded cranks is to underestimate the danger they pose. The time has come to put the creation science movement where it belongs — and that is definitely not in the science classroom.

Sources

The following notes refer to quotations from other sources cited in the text. All those attributed to Dr. Henry M. Morris are taken from his book *Scientific Creationism* (General Edition), San Diego: Creation-Life Publishers, 1974. All those attributed to Dr. Duane T. Gish are taken from his book *Evolution: The Fossils Say No!,* San Diego: Creation-Life Publishers, 1973. Complete information on all other sources cited here is to be found in the References which begin on page 194.

CHAPTER ONE
PAGE
 6   Morris, p. 21
 11   Morris, pp. 43, 18
 13   Morris, pp. 92, 97, 111-12
 14   Morris, p. 10

CHAPTER THREE
 30   Gish, pp. 3, 17
 32   Morris, pp. 180-1

CHAPTER FIVE
 54   Morris, p. 252
 55   paraphrased from Morris, pp. 252-3
 57   Morris, p. 252
 58   paraphrased from Morris, pp. 118-19
 66   Morris, p. 119

CHAPTER SIX
 68   Morris, p. 79
 71   Morris, p. 78
 74   Gish, p. 47
 74   Morris, pp. 81-2

CHAPTER SEVEN
 79   Morris, p. 99
 85   Morris, p. 18
 87   Pettersson, p. 132

CHAPTER EIGHT
 91   Morris, pp. 97, 99, 97

CHAPTER NINE
 102   Gish, p. 45
 102   paraphrased from Morris, p. 79
 105   Gish, p. 45
 107   Morris, pp. 121-2, 122

CHAPTER TEN
116   Morris, p. 85
116   Gish, p. 60
121   Gish, p. 62
123   Gish, p. 62. In addition
       to the clawed wings,
       Dr. Gish refers to a
       second feature of the
       Hoatzin that has been
       "used to impute a
       reptilian ancestry to
       Archaeopteryx," name-
       ly poor flying ability.
       As we have no way
       of knowing the flying
       ability of *Archaeopteryx*,
       this is irrelevant to
       the discussion and
       has been accordingly
       omitted from the text.
       (Dr. Gish also men-
       tions that the Hoatzin
       has a small sternum,
       but there is no evi-
       dence of a sternum
       in *Archaeopteryx*.)
125   Gish, p. 60

CHAPTER ELEVEN
127   Morris, p. 83
140   Romer, p. 17
140   Morris, p. 83
140   Gish, p. 63

CHAPTER TWELVE
147   Ewart, p. 343

CHAPTER THIRTEEN
151   Owen, pp. 133-4
154   Gish pp. 54, 55
155   Gish, p. 55
174   Dart, 1925; reprinted
       in Leakey and Prost,
       1971, p. 208
176   Gish, p. 82
176   Leakey, pp. 245, 96
184   Morris, p. 174
185   Morris, p. 178

Picture Credits

7, 9, 10, 35, 73, 75   Jay Sobel
82   J. H. McAndrews, Department of
Botany, Royal Ontario Museum
100   From G. A. Mantell. 1854.
Medals of Creation. London: Henry
G. Bohn.
101   Anker Odum, Art Department,
Exhibit Design Services, R.O.M.
112   Bill Robertson, Arlene Reiss,
Photography Department and
Department of Vertebrate
Palaeontology, R.O.M.
113   (a) From O. C. Marsh. 1896.
The dinosaurs of North America.
*Sixth Annual Report of the United
States Geological Survey.* Washington:
Government Printing Office.

(b) Julian Mulock (c) From Marsh
118   Chris McGowan
120   Anker Odum, R.O.M.
124   Bill Robertson, Arlene Reiss,
R.O.M.
128, 129, 131, 132, 136   Anker Odum,
R.O.M.
144   Jay Sobel, redrawn from
A. S. Romer. 1966. *Vertebrate Paleontology.* Chicago: University of
Chicago Press.
148   Jeff Thomason, Department of
Vertebrate Palaeontology, R.O.M.
153, 155   Jay Sobel, redrawn from
Romer.
165, 166, 167, 168, 169, 172, 173   Bill
Robertson, R.O.M.

References

CHAPTER 1

Gish, D. T. 1972. *Evolution: the fossils say no!* San Diego, Calif.: Creation-Life Publishers.

Holmes, A., and Holmes, D. C. 1978. *Holmes principles of physical geology.* Sunbury-on-Thames: Thomas Nelson.

King, M. C., and Wilson, A. C. 1975. Evolution at two levels in humans and chimpanzees. *Science* 188: 107–18.

Lyell, C. 1853. *Principles of geology.* 9th ed. Boston: Little, Brown.

Morris, H. M. 1974. *Scientific Creationism.* San Diego, Calif.: Creation-Life Publishers.

Rudwick, M. J. S. 1972. *The meaning of fossils: episodes in the history of palaeontology.* New York: Science History Publications.

Weidner, R. T., and Sells, R. L. 1975. *Elementary physics.* Boston: Allyn and Bacon.

CHAPTER 2

Darwin, C. 1839. *Journal of researches into the geology and natural history of the various countries visited by H.M.S. Beagle.* London: Henry Colbourn.

_____. 1859. *The origin of species by means of natural selection or the preservation of favoured races in the struggle for life.* London: John Murray.

Darwin, F., ed. 1887. *The life and letters of Charles Darwin*. London: John Murray.

Rudwick, M. J. S. 1972. *The meaning of fossils: episodes in the history of palaeontology*. New York: Science History Publications.

CHAPTER 3

Berry, R. J., and Jakobson, M. E. 1975. Ecological genetics of an island population of the house mouse (*Mus musculus*). *Journal of Zoology* 175: 532–40.

––––––, and Jakobson, M. E. 1975. Adaptation and adaptability in wild-living house mice (*Mus musculus*). *Journal of Zoology* 176: 391–402.

Eldredge, N., and Gould, S. J. 1972. Punctuated equilibrium: an alternative to phyletic gradualism. In *Models in paleobiology*, ed. T. J. M. Shopf. San Francisco: Freeman and Cooper.

Greenwood, P. H. 1974. The cichlid fishes of Lake Victoria, East Africa: the biology and evolution of a species flock. *Bulletin of the British Museum (Natural History)*, Zoology, Supplement 6.

Kurten, B. 1968. *Pleistocene mammals of Europe*. London: Weidenfeld and Nicolson.

McNeilly, T., and Bradshaw, A. D. 1968. Evolutionary processes in populations of copper tolerant *Agrostis tenuis* Sibth. *Evolution* 22: 108–18.

CHAPTER 4

A series of accounts of the Arkansas trial appeared in the "News and Comment" section of the journal *Science*.
See *Science* (1981) 214: 1101–4, 1224; (1982) 215: 142–6.

Anders, E., Hayatso, R., and Studier, M. H. 1973. Organic compounds in meteorites. *Science* 182: 781–90.

Dickerson, R. E. 1978. Chemical evolution and the origin of life. *Scientific American* 239: 70–86.

Hayatsu, R. 1964. Orgueil meteorite: Organic nitrogen contents. *Science* 146: 1291–2.

Stoks, P. G., and Schwarts, A. W. 1978. Uracil in carbonaceous meteorites. *Nature* 282: 709–10.

CHAPTER 5

Schaller, G. B. 1972. *The Serengeti Lion*. Chicago: University of Chicago Press.

Sikes, S. K. 1971. *The natural history of the African Elephant*. London: Weidenfeld and Nicolson.
Stebbins, R. C. 1954. *Amphibians and reptiles of western North America*. New York: McGraw-Hill.
Tyler, M. J. 1976. *Frogs*. Sydney: Collins.

CHAPTER 6

Beklemishev, W. N. 1969. *Principles of comparative anatomy of invertebrates*. Edinburgh: Oliver and Boyd.
Borradale, L. A., et al. 1961. *The Invertebrata*. Cambridge: Cambridge University Press.
Cloney, R. A. 1978. Ascidian metamorphosis: review and analysis. In *Settlement and metamorphosis of marine invertebrate larvae*, ed. F. S. Chia and M. E. Rice. New York: Elsevier.
Kudo, R. 1954. *Protozoology*. Springfield, Ill.: Charles C. Thomas.
Plough, H. H. 1978. *Sea squirts of the Atlantic Continental Shelf from Maine to Texas*. Baltimore: Johns Hopkins University Press.

CHAPTER 7

Abell, G. 1969. *Exploration of the Universe*. New York: Holt, Rinehart and Winston.
Berry, R. J. 1977. *Inheritance and natural history*. London: Collins.
Continents adrift. 1970. *Readings from Scientific American*. San Francisco: W. H. Freeman.
Hallam, A. 1973. *A revolution in the Earth sciences*. Oxford: Clarendon Press.
Kron, R. G. 1982. The most distant known galaxies. *Science* 216: 265–9.
Lack, D. 1954. *The natural regulation of animal numbers*. Oxford: Clarendon Press.
McAndrews, J. H. 1976. Fossil history of man's impact on the Canadian flora: an example from southern Ontario. *Canadian Botanical Association Bulletin* 9: 1–6.
Morrison, D. 1982. Astronomy and creationism. *Mercury* September-October: 144–7.
Murton, R. K., and Westwood, N. J. 1977. *Avian breeding cycles*. Oxford: Clarendon Press.
Pettersson, H. 1960. Cosmic spherules and meteoric dust. *Scientific American* 202: 123–32.
Worzel, J. L., et al. 1973. *Initial Reports of the Deep Sea Drilling Project*, volume 10. Washington: U. S. Government Printing Office.

York, D., and Farquar, R. M. 1972. *The Earth's age and geo-chronology*. Oxford: Pergamon Press.

CHAPTER 8

Galdikas, B. M. F. 1978. Orangutan death and scavenging by pigs. *Science* 200: 68–70.

McGowan, C. 1983. *The successful dragons: a natural history of extinct reptiles*. Toronto: Samuel-Stevens.

Oliver, W. R. B. 1949. The Moas of New Zealand and Australia. *Bulletin of the Dominion Museum*, no. 15.

Romer, A. S. 1970. The Chañares (Argentina) Triassic reptile fauna VI. A chiniquodontid cynodont with an incipient squamosal-dentary jaw articulation. *Breviora*, no. 344:1–18.

CHAPTER 9

Black, R. M. 1973. *The elements of palaeontology*. London: Cambridge University Press.

Cloud, P., and Glaessner, M. F. 1982. The Ediacarian Period and System: Metazoa inherit the Earth. *Science* 217: 783–92.

Cowie, J. W. 1967. Life in Pre-Cambrian and early Cambrian times. In *The Fossil Record*, pp. 17–35. Geological Society of London.

Edwards, D. 1980. Early land floras. In *The terrestrial environment and the origin of land vertebrates*, ed. A. L. Panchen. London: Academic Press.

Ford, T. D. 1979. Precambrian fossils and the origin of the phyla. In *The origin of the major invertebrate groups*, ed. M. R. House. Systematics Association Publication No. 12.

————. 1980. The Ediacarian fossils of Charnwood Forest, Leicestershire. *Proceedings of the Geologists' Association* 91:81–3.

Glaessner, M. F., and Wade, M. 1966. The late Precambrian fossils from Ediacara, South Australia. *Palaeontology* 9: 599–628.

Milne, D. H. 1981. How to debate with creationists — and "win." *American biology teacher* 43: 235–45.

Rolf, W. D. I. 1980. Early invertebrate terrestrial faunas. In *The terrestrial environment and the origin of land vertebrates*, ed. A. L. Panchen. London: Academic Press.

Romer, A. S. 1966. *Vertebrate palaeontology*. Chicago: University of Chicago Press.

Schopf, J. W. 1978. The evolution of the earliest cells. *Scientific American* 239: 110–35.

Von Zittel, K. 1901. *History of geology and paleontology.* London: Walter Scott.

Data for first appearances of fossil groups from:

Andrews, H. N. 1961. *Studies in palaeobotany.* New York: Jack Wiley.

Harland, W. B., et al., eds. 1967. *The fossil record.* London: Geological Society.

CHAPTER 10

Gould, S. J., and Vrba, E. S. 1982. Exaptation — a missing term in the science of form. *Paleobiology.* 8:4–15.

Heilmann, G. 1926. *The origin of birds.* London: H. F. and G. Whitherby.

Huxley, T. H. 1868. On the animals which are most nearly intermediate between the birds and reptiles. *The Annals and Magazine of Natural History* 2: 66–75. London.

King, M. C., and Wilson, A. C. 1975. Evolution at two levels in humans and chimpanzees. *Science* 188: 107–16.

Kollar, E. J., and Fisher, C. 1980. Tooth induction on chick epithelium: expression of quiescent genes for enamel synthesis. *Science* 207: 993–5.

Ostrom, J. H. 1970. *Archaeopteryx*, notice of a "new" specimen. *Science* 170: 537–8.

_____. 1974. *Archaeopteryx* and the origin of flight. *Quarterly Review of Biology* 49: 27–47.

_____. 1976. *Archaeopteryx* and the origin of birds. *Biological Journal of the Linnean Society* 8: 91–182.

Owen, R. 1864. On the *Archaeopteryx* of von Meyer, with a description of the fossil remains of a long-tailed species, from the lithographic stone of Solenhofen. *Philosophical Transactions of the Royal Society of London* 153: 33–47.

Regal, P. J. 1975. The evolutionary origin of feathers. *Quarterly Review of Biology* 50: 35–66.

Wagner, J. A. 1861. Ueber ein neues Augenblick mit Vogelfedern versehenes Reptil aus dem Solenhofener lithographischen Schiefer. *Sitzungsberichte bayerische Akademie Wissenschaften* 2: 146–54.

CHAPTER 11

Clemens, W. A. 1970. Mesozoic mammalian evolution. *Annual Review of Ecology and Systematics* 1: 357–90.

Crompton, A. W., and Jenkins, F. A., Jr. 1973. Mammals from reptiles: A review of mammalian origins. *Annual Review of Earth and Planetary Sciences* 1:131–55.

_____, and Jenkins, F. A., Jr. 1978. Mesozoic mammals. In *Evolution of African mammals*, ed. V. J. Maglio and H. B. S. Cooke. Cambridge, Mass.: Harvard University Press.

Grant, T. Q., and Dawson, T. J. 1978. Temperature regulation in the platypus *Ornithorhynchus anatinus*: Production and loss of metabolic heat in air and water. *Physiological Zoology* 51: 315–32.

Hopson, J. A. 1976. Mammal-like reptiles and the origin of mammals. *Discovery* 2: 25–33.

Jenkins, F. A., Jr. 1970. The Chañares (Argentina) Triassic reptile fauna VII. The postcranial skeleton of the traversodontid *Massetognathus pascuali*. (Therapsida, Cynodontia). *Breviora*, no. 352: 1–28.

Romer, A. S. 1969. Topics in therapsid evolution and classification. *Bulletin of the Indian Geologists' Association* 2: 15–26.

_____. 1969. The Chañares (Argentina) Triassic reptile fauna V. A new chiniquodontid cynodont, *Probolesodon lewisi* — Cynodont ancestry. *Breviora*, no. 333: 1–24.

_____. 1970. The Chañares (Argentina) Triassic reptile fauna VI. A chiniquodontid cynodont with an incipient squamosal-dentary jaw articulation. *Breviora*, no. 344: 1–18.

CHAPTER 12

Camp, C. L., and Smith, N. 1942. Phylogeny and function of the digital ligaments of the horse. *Memoirs of the University of California, Berkeley*, no. 13: 69–124.

Ewart, J. C. 1894. The development of the skeleton of the limbs of the horse, with observations on polydactyly. *Journal of Anatomy and Physiology* 28: 342–69.

MacFadden, B. J. 1976. Cladistic analysis of primitive equids, with notes on other perissodactyls. *Systematic Zoology* 25: 1–15.

Simpson, G. G. 1951. *Horses*. New York: Oxford University Press.

Struthers, J. 1893. On the development of the bones of the foot of the horse, and of digital bones generally and on a case of polydactyly in the horse. *Journal of Anatomy and Physiology* 28: 51–62.

CHAPTER 13

Bentley, P. J. 1966. Adaptations of amphibia to arid environments. *Science* 152: 619–23.

Carroll, R. C. 1981. Plesiosaur ancestors from the Upper Permian of Madagascar. *Philosophical Transactions of the Royal Society of London* 293: 315–83.

Noble, G. K. 1931. *Biology of the Amphibia.* New York: McGraw-Hill.

Owen, R. 1841. Description of the *Lepidosiren annectens. Transactions of the Linnean Society of London* 18: 327–61.

_____ . 1859. Palaeontology. In *Encyclopaedia Britannica*, vol. 17, pp. 91–176.

Patterson, C. 1980. Origin of tetrapods: Historical introduction of the problem. In *The terrestrial environment and the origin of land vertebrates*, ed. A. L. Panchen. London: Academic Press.

Porter, K. R. 1972. *Herpetology.* London: W. B. Saunders.

Rackoff, J. S. 1980. The origin of the tetrapod limb and the ancestry of tetrapods. In *The terrestrial environment and the origin of land vertebrates*, ed. A. L. Panchen. London: Academic Press.

Thompson, K. S. 1980. The ecology of Devonian lobe-finned fishes. In *The terrestrial environment and the origin of land vertebrates*, ed. A. L. Panchen. London: Academic Press.

Tyler, M. J. 1976. *Frogs.* Sydney: Collins.

CHAPTER 14

Brace, C. L., Nelson, H., Kom N., and Brace, M. L. 1979. *Atlas of human evolution.* New York: Holt, Rinehart and Winston.

Cole, S. 1975. *Leakey's Luck.* London: Collins.

Crile, G., and Quiring, D. P. 1940. A record of the body weight and certain organs and gland weight of 3690 animals. *Ohio Journal of Science* 40: 219–59.

Elliot, D. G. 1912. *A review of the primates.* New York: American Museum of Natural History.

Jerison, H. J. 1973. *Evolution of the brain and intelligence.* New York: Academic Press.

Johanson, D. C., and White, T. D. 1979. A systematic assessment of early African hominids. *Science* 203: 320–30.

_____ , and Edey, M. A. 1981. *Lucy: the beginnings of humankind.* New York: Warner Books.

Leakey, L. S. B., Prost, J., and Prost, S., eds. 1971. *Adam or ape: a sourcebook of discoveries about early man.* Cambridge, Mass.: Schenkman Publishing Company.

Leakey, R. E. 1971. Further evidence of Lower Pleistocene hominids from East Rudolf, North Kenya. *Nature* 231: 241–5.

_____ . 1976. Hominids in Africa. *American Scientist* 64: 174–8.

_____ , and Lewin, R. 1977. *Origins.* New York: E. P. Dutton.

Le Gros Clark, W. E. 1950. Hominid characters of the australopithecine dentition. *Journal of the Royal Anthropological Institute of Great Britain and Ireland* 80: 37–53.

———. 1959. *The antecedents of man*. New York: Harper and Row.
Pilbeam, D. 1970. *Gigantopithecus* and the origins of Hominidae. *Nature* 225: 516–19.

The encephalization quotient, $EQ = \dfrac{\text{calculated brain weight}}{\text{actual brain weight}}$

The brain weight of a mammal is calculated from the equation
brain weight $= 0.12 \times (\text{body weight})^{2/3}$
This equation was determined empirically from data obtained from a large sample of mammals. For further information, see Jerison 1973.

Index